LECTURES ON
Boolean
Algebras

PAUL R. HALMOS

DOVER PUBLICATIONS, INC.
Mineola, New York

Bibliographical Note

This Dover edition, first published in 2018, is an unabridged republication of the work originally published in the Van Nostrand Reinhold Mathematical Studies series by Van Nostrand Reinhold Company, London, in 1963.

Library of Congress Cataloging-in-Publication Data

Names: Halmos, Paul R. (Paul Richard), 1916–2006, author.
Title: Lectures on Boolean algebras / Paul R. Halmos.
Other titles: Boolean algebras
Description: Dover edition. | Mineola, New York : Dover Publications, Inc., 2018. | Originally published: London, England ; Princeton, N.J. : Van Nostrand, 1963. | Includes index.
Identifiers: LCCN 2018010158| ISBN 9780486828046 | ISBN 0486828042
Subjects: LCSH: Algebra, Boolean.
Classification: LCC QA10.3 .H34 2018 | DDC 511.3/24—dc23
LC record available at https://lccn.loc.gov/2018010158

Manufactured in the United States by LSC Communications
82804201   2018
www.doverpublications.com

Contents

Preface

IN 1959 I lectured on Boolean algebras at the University of Chicago. A mimeographed version of the notes on which the lectures were based circulated for about two years; this volume contains those notes, corrected and revised. Most of the corrections were suggested by Peter Crawley. To judge by his detailed and precise suggestions, he must have read every word, checked every reference, and weighed every argument, and I am very grateful to him for his help. This is not to say that he is to be held responsible for the imperfections that remain, and, in particular, I alone am responsible for all expressions of personal opinion and irreverent viewpoint.

P. R. H.

Ann Arbor, Michigan
January, 1963

§ 1. Boolean Rings

An element p of a ring is *idempotent* if $p^2 = p$. A *Boolean ring* is a ring with unit in which every element is idempotent. Warning: a ring with unit is by definition a ring with a distinguished element 1 that acts as a multiplicative identity and that is distinct from the additive identity 0. The effect of the last proviso is to exclude from consideration the trivial ring consisting of 0 alone. The phrase "with unit" is sometimes omitted from the definition of a Boolean ring; in that case our present concept is called a "Boolean ring with unit."

Every Boolean ring contains 0 and 1; the simplest Boolean ring contains nothing else. Indeed, the ring of integers modulo 2 is a Boolean ring. This particular Boolean ring will be denoted throughout by the same symbol as the ordinary integer 2. The notation is not commonly used, but it is very convenient. It is in accordance with von Neumann's definition of ordinal number, with sound general principles of notational economy, and (in logical expressions such as "two-valued") with idiomatic linguistic usage.

A non-trivial and natural example of a Boolean ring is the set 2^X of all functions from an arbitrary non-empty set X into 2. The elements of 2^X will be called 2-valued functions on X. The distinguished elements and operations in 2^X are defined pointwise. This means that 0 and 1 in 2^X are the functions defined, for each x in X, by

$$0(x) = 0 \quad \text{and} \quad 1(x) = 1,$$

1

and, if p and q are 2-valued functions on X, then the functions $p + q$ and pq are defined by

$$(p + q)(x) = p(x) + q(x) \text{ and } (pq)(x) = p(x)q(x).$$

These equations make sense; their right sides refer to elements of 2. The assumption that $X \neq \emptyset$ is needed to guarantee that 0 and 1 are distinct.

For another example of a Boolean ring let A be the set of all idempotent elements in a commutative (!) ring R with unit, with addition redefined so that the new sum of p and q in A is $p + q - 2pq$. The distinguished elements of A are the same as those of R, and multiplication in A is just the restriction of multiplication in R. The verification that A becomes a Boolean ring this way is an amusing exercise in ring axiomatics. Commutativity is used repeatedly; it is needed, for instance, to prove that A is closed under multiplication.

The main condition in the definition of a Boolean ring (idempotence) has quite a strong influence on the structure of such rings. Two of its most surprising consequences are that (1) a Boolean ring A has *characteristic* 2 (that is, $p + p = 0$ for every p in A), and (2) a Boolean ring is commutative. For the proof, compute $(p + q)^2$, and use idempotence to conclude that $pq + qp = 0$. This result implies the two assertions, one after another, as follows. Put $p = q$ and use idempotence to get (1); since (1) implies that every element is equal to its own negative, the fact that $pq = -pq$ yields (2).

Since, as we now know, subtraction in Boolean rings is the same as addition, it is never necessary to use the minus sign for additive inverses, and we shall never again do so. A little later we shall meet another natural use for it.

(1) Prove that every Boolean ring without a unit can be embedded into a Boolean ring with a unit. To what extent is this extension procedure unique?

(2) Can every Boolean ring with unit be obtained by adjoining a unit to a Boolean ring without a unit?

(3) A *Boolean group* is an (additive) abelian group in which every element has order two (that is, $p + p = 0$ for all p). Is every Boolean group the additive group of some Boolean ring?

§ 2. Boolean algebras

Let X be an arbitrary non-empty set and let $\mathcal{P}(X)$ (the *power set* of X) be the class of all subsets of X. There is a way of introducing a Boolean structure into $\mathcal{P}(X)$, as follows. The distinguished elements are defined by

$$0 = \emptyset \text{ and } 1 = X,$$

and, if P and Q are subsets of X, then, by definition,

$$P + Q = (P \cap Q') \cup (P' \cap Q) \text{ and } PQ = P \cap Q.$$

The symbols $\cup$, $\cap$, and $'$ refer, of course, to the ordinary set-theoretic concepts of union, intersection, and complement. The easiest way to verify that the result is indeed a Boolean ring is to establish a one-to-one correspondence between $\mathcal{P}(X)$ and 2^X so that the elements and operations here defined correspond exactly to the distinguished elements and operations of 2^X. The 2-valued function p corresponding to a subset P of X is just its *characteristic function,* that is, the function defined for each x in X by

$$p(x) = \begin{cases} 1 \text{ if } x \in P, \\\\ 0 \text{ if } x \in' p. \end{cases}$$

Observe that the Boolean sum $P + Q$ is what is usually known in set theory as the *symmetric difference* of P and Q.

Motivated by this set-theoretic example, we can introduce into every Boolean ring operations very much like the set-theoretic ones; all we need to do is to write

(1) $$p \wedge q = pq,$$

(2) $$p \vee q = p + q + pq,$$

(3) $$p' = 1 + p.$$

Meet, join, and *complement,* respectively, are among the several possible widely adopted names of these operations. It should come as no surprise that plus and times can be recaptured from meet, join, and complement; indeed

(4) $$pq = p \wedge q,$$

(5) $$p + q = (p \wedge q') \vee (p' \wedge q).$$

From this it follows that it must be possible to use meet, join, and complement (and, of course, 0 and 1) as the primitive terms of an axiomatization of Boolean rings, and, indeed, this can be done in many ways.

In principle the task is an easy one. All we have to do is express each of the defining conditions (axioms) of a Boolean ring in terms of meet, join, and complement, and then use the resulting conditions, or some others strong enough to imply them, as axioms. Here is a wastefully large set of conditions, more than strong enough for the purpose.

(6) $\qquad 0' = 1 \qquad\qquad\qquad\qquad 1' = 0$

(7) $\qquad p \wedge 0 = 0 \qquad\qquad\qquad p \vee 1 = 1$

(8) $\qquad p \wedge 1 = p \qquad\qquad\qquad p \vee 0 = p$

(9) $\qquad p \wedge p' = 0 \qquad\qquad\qquad p \vee p' = 1$

(10) $\qquad\qquad\qquad p'' = p$

(11) $\qquad p \wedge p = p \qquad\qquad\qquad p \vee p = p$

(12) $\quad (p \wedge q)' = p' \vee q' \qquad\qquad (p \vee q)' = p' \wedge q'$

(13) $\qquad p \wedge q = q \wedge p \qquad\qquad\qquad p \vee q = q \vee p$

(14) $\quad p \wedge (q \wedge r) = (p \wedge q) \wedge r \qquad p \vee (q \vee r) = (p \vee q) \vee r$

(15) $\; p \wedge (q \vee r) = (p \wedge q) \vee (p \wedge r) \quad p \vee (q \wedge r) = (p \vee q) \wedge (p \vee r)$

The problem of selecting small subsets of this set of conditions that are strong enough to imply them all is one of dull axiomatics. For the sake of the record: one solution of the problem is given by the pairs of conditions (8), (9), the commutative laws (13), and the distributive laws (15). To prove that these four pairs imply all the other conditions, and, in particular, to prove that they imply the De Morgan laws (12) and the associative laws (14), involves some non-trivial trickery.

The customary succinct way of summarizing the preceding discussion motivates the following definition. Let us call a *Boolean algebra* a set A together with two distinct distinguished elements 0 and 1, two binary operations $\wedge$ and $\vee$, and a unary operation $'$, satisfying the identities (6)-(15).

The succinct summary says that every Boolean ring is a Boolean algebra, and vice versa. A somewhat more precise statement is somewhat clumsier. It says that if the Boolean operations (meet, join, and complement) are defined in a Boolean ring A by (1)-(3), then A becomes a Boolean algebra; and, backwards, if the ring operations are defined in a Boolean algebra A by (4) and (5), then A becomes a Boolean ring. In these notes we shall use the two terms (Boolean ring and Boolean algebra) almost as if they were synonymous, selecting on each occasion the one that seems intuitively more appropriate. Since our motivation comes from set theory, we shall speak of Boolean algebras much more often than of Boolean rings.

Here is a comment on notation, inspired by the associative laws (14). It is an elementary consequence of those laws that if $p_1, \cdots, p_n$ are elements of a Boolean algebra, then $p_1 \vee \cdots \vee p_n$ makes sense. The point is, of course, that since such joins are independent of how they are bracketed, it is not necessary to indicate any bracketing at all. The element $p_1 \vee \cdots \vee p_n$ may alternatively be denoted by $\bigvee_{i=1}^{n} p_i$, or, in case no confusion is possible, simply by $\bigvee_i p_i$.

If we make simultaneous use of both the commutative and the associative laws, we can derive a slight but useful generalization of the preceding comment. If E is a non-empty finite subset of a Boolean algebra, then the set E has a uniquely determined join, independent of any order or bracketing that may be used in writing it down. (In case E is a singleton, it is natural to identify that join with the unique element in E.) We shall denote the join of E by $\bigvee E$.

Both the preceding comments apply to meets as well as to joins. The corresponding symbols are, of course,

$\bigwedge_{i=1}^{n} p_i$, or $\bigwedge_i p_i$, and $\bigwedge E$.

The point of view of Boolean algebras makes it possible to give a simple and natural description of an example that would be quite awkward to treat from the point of view of Boolean rings. Let m be an integer greater than 1, and let A be the set of all positive integral divisors of m. Define the Boolean structure of A by the equations

$$0 = 1,$$

$$1 = m,$$

$$p \wedge q = \text{g. c. d. } \{p, q\},$$

$$p \vee q = \text{l. c. m. } \{p, q\},$$

$$p' = m/p.$$

It turns out that, with the distinguished elements and operations so defined, A forms a Boolean algebra if and only if m is square-free (that is, m is not divisible by the square of any prime). Query: what are the number-theoretic expressions of the ring operations in this Boolean algebra? And, while we are on the subject, what are the expressions for the Boolean operations in the Boolean ring A consisting of the idempotent elements of an arbitrary commutative ring R with unit? (See §1.) The answer to this question is slightly different from (1)-(3); those equations give the answer in terms of the ring operations in A, and what is wanted is an answer in terms of the ring operations in R.

A technical reason for preferring the language of Boolean algebras to that of Boolean rings is the so-called *principle of duality*. The principle consists in observing that if 0 and 1 are interchanged in the identities (6)-(15), and if, at the

same time, $\wedge$ and $\vee$ are interchanged, then those identities
are merely permuted among themselves. It follows that the
same is true for all the consequences of those identities.
The general theorems about Boolean algebras, and, for that
matter, their proofs also, come in dual pairs. A practical
consequence of this principle, often exploited in what
follows, is that in the theory of Boolean algebras it is
sufficient to state and to prove only half the theorems; the
other half come gratis from the principle of duality.

A slight misunderstanding can arise about the meaning
of duality, and often does. It is well worth while to clear it
up once and for all, especially since the clarification is
quite amusing in its own right. If an experienced Boolean
algebraist is asked for the dual of a Boolean polynomial,
such as say $p \vee q$, his answer might be $p \wedge q$ one day and
$p' \wedge q'$ another day; the answer $p' \vee q'$ is less likely but
not impossible. (The definition of Boolean polynomials is
the same as that of ordinary polynomials, except that the
admissible operations are not addition and multiplication
but meet, join, and complement.) What is needed here is
some careful terminological distinction. Let us restrict
attention to the completely typical case of a polynomial
$f(p, q)$ in two variables. The *complement* of $f(p, q)$ is by
definition $(f(p, q))'$, abbreviated $f'(p, q)$; the *dual* of $f(p, q)$
is $f'(p', q')$; the *contradual* of $f(p, q)$ is $f(p', q')$. What
goes on here is that the group acting, in an obvious way, is
not the group of order two, but the Klein four-group. This
comment was made by Gottschalk (J.S.L., vol. 18) who
describes the situation by speaking of the principle of
quaternality.

A word of warning: the word "duality" is frequently used
in contexts startlingly different from each other and from the
one we met above. This is true even within the theory of

Boolean algebras, where, for instance, a topological duality theory turns out to play a much more important role than the elementary algebraic one. If the context alone is not sufficient to indicate the intended meaning, great care must be exercised to avoid confusion.

Exercises

(1) The four pairs of identities (8), (9), (13), and (15) constitute a set of axioms for Boolean algebras; are they an independent set?

(2) Prove that the following identities constitute a set of axioms for Boolean algebras: $\vee$ is commutative and associative, and $(p' \vee q')' \vee (p' \vee q)' = p$.

(3) Prove that the following identities constitute a set of axioms for Boolean algebras: $p'' = p$, $p \vee (q \vee q') = p$, and $p \vee (q \vee r)' = ((q' \vee p)' \vee (r' \vee p)')'$.

(4) Prove that the commutative and associative laws for $\vee$, together with the requirement that (for all p, q, and r) $p \vee q' = r \vee r'$ if and only if $p \vee q = p$, constitute a set of axioms for Boolean algebras.

§ 3. Fields of sets

To form $\mathcal{P}(X)$ is not the only natural way to make a Boolean algebra out of a non-empty set X. A more general way is to consider an arbitrary non-empty subclass A of $\mathcal{P}(X)$ such that if P and Q are in A, then $P \cap Q$, $P \cup Q$, and P' are also in A. Since A contains at least one element, it follows that A contains $\emptyset$ and X (cf. (2.9)), and hence that A is a Boolean algebra. Every Boolean algebra obtained in this way is called a *field* (of sets). There is usually no danger in denoting a field of sets by the same symbol as the class of sets that go to make it up. This does not, however, justify the conclusion (it is false) that set-theoretic intersection, union, and complement are the only possible operations that convert a class of sets into a Boolean algebra.

A subset P of a set X is *cofinite* (in X) if its complement P' is finite. The class A of all those subsets of a non-empty set X that are either finite or cofinite is a field of subsets of X. If X itself is finite, then A is simply $\mathcal{P}(X)$; if X is infinite, then A is a new example of a Boolean algebra.

The preceding construction can be generalized. Call a subset P of X *cocountable* (in X) if its complement X' is countable. The class of all those subsets of X that are either countable or cocountable is a field of subsets of X. Different description of the same field: the class of all those subsets P of X for which the cardinal number of either P of P' is less than or equal to $\aleph_0$. A further generalization is obtained by using an arbitrary cardinal number in place of $\aleph_0$.

Let X be the set of all integers (positive, negative, or zero), and let m be an arbitrary integer. A subset P of X is *periodic* of *period m* if it coincides with the set obtained by adding m to each of its elements. The class A of all periodic sets of period m is a field of subsets of X. If $m = 0$, then A is simply $\mathcal{P}(X)$. If $m = 1$, then A consists of just the two sets $\emptyset$ and X. In all other cases A is a new example of a Boolean algebra.

Let X be the set of all real numbers. A *left half-closed interval* (or, for brevity, since this is the only kind we shall consider, *a half-closed interval*) is a set of one of the forms $(-\infty, b)$, or $[a, b)$, or $[a, +\infty)$, that is, the set of all those elements x in X for which $x < b$, or $a \leq x < b$, or $a \leq x$, where, of course, a and b themselves are real numbers and $a < b$. The class A of all finite unions of half-closed intervals is a field of subsets of X. A useful variant of this example uses the closed unit interval $[0, 1]$ in the role of X. In this case it is convenient to stretch the terminology so as to include the closed intervals $[a, 1]$ and the degenerate interval $\{1\}$ among half-closed intervals.

Valuable examples of fields of sets can be defined in the unit square, as follows. Call a subset P of the square *vertical* if, along with each point in P, every point of the vertical segment through that point also belongs to P. In other words, P is vertical if the presence of (x_0, y_0) in P implies the presence in P of (x_0, y) for every y in $[0, 1]$. If A is any field of subsets of the square, then the class of all vertical sets in A is another, and, in particular, the class of all vertical sets is a field of sets. Here are two comments that are trivial but sometimes useful: (1) the *horizontal* sets (whose definition may safely be left to the reader) constitute just as good a field as the vertical sets, and (2) the Cartesian product of any two non-empty sets is, for these purposes, just as good as the unit square.

Other examples of fields of subsets of the real line (or of an interval in the line) are given by the class of all Lebesgue measurable sets and by the class of all Borel sets. These examples are readily generalized to arbitrary measure spaces and to arbitrary topological spaces, respectively.

A subset P of a topological space X may be simultaneously closed and open; consider, for example, the sets $\emptyset$
and X. A set with this property is called *clopen*. If the
space is connected, then $\emptyset$ and X are the only clopen sets,
and conversely; this, in fact, constitutes a possible definition of connectedness. In any case, the class of all clopen
subsets of a topological space X, connected or not, is a
field of subsets of X.

Our last example, for now, depends on the concept of
the boundary of a subset P of a topological space X; recall
that the boundary of P is, by definition, the intersection
$P^- \cap P'^-$, where, for typographical convenience, P^- (not
$\bar{P}$) denotes the closure of P. Assertion: if "small", when
applied to subsets of X is interpreted in any reasonable way,
then the class of all sets with small boundaries is a field.
One reasonable interpretation of "small" is "countable";
another is "nowhere dense". Recall that a set P is
nowhere dense if the interior of its closure is empty. The
proof of the assertion depends on the easily available
topological fact that the boundary of the union of two sets
is included in the union of their boundaries.

Exercises

(1) If m is a positive integer, and if A is the class of
all those sets of integers that are periodic of some period
greater than m, is A a field of sets?

§ 4. Regular open sets

The purpose of this section is to discuss one more
example of a Boolean algebra. This example, the most
intricate of all the ones so far, is one in which the elements

of the Boolean algebra are subsets of a set, but the operations are not the usual set-theoretic ones, so that the Boolean algebra is not a field of sets. Artificial examples of this kind are not hard to manufacture; the example that follows arises rather naturally and plays an important role in the general theory of Boolean algebras.

Let X be an arbitrary (non-empty) topological space. Recall that an open set in X is called *regular* in case it coincides with the interior of its own closure. In other words, since the interior of a set P is $P^{'-'}$, the set P is regular if and only if $P = P^{-'-'}$. It is convenient, in this connection, to write $P^{\perp} = P^{-'}$; in these terms P is regular if and only if $P = P^{\perp\perp}$. Note incidentally that a set P is open (nothing is said about regularity here) if and only if it has the form $Q^{\perp}$ for some set Q. Indeed, if $P = Q^{\perp}$, then P is the complement of the closed set Q^{-}, and, conversely, if P is open, then $P = Q^{\perp}$ where Q is the complement of P.

THEOREM 1. *The class A of all regular open sets of a non-empty topological space X is a Boolean algebra with respect to the distinguished Boolean elements and operations defined by*

(1) $$0 = \varnothing,$$

(2) $$1 = X,$$

(3) $$P \wedge Q = P \cap Q,$$

(4) $$P \vee Q = (P \cup Q)^{\perp\perp};$$

(5) $$P' = P^{\perp},$$

where $P^{\perp}$, for every set P, is the complement of the closure of P, and where $P^{\perp\perp}$, of course, is $(P^{\perp})^{\perp}$.

The proof of the theorem depends on several small lemmas of some independent interest. Observe that the first thing to prove is that the right sides of (1)-(5) are regular open sets. For (1) and (2) this is obvious, but for (3), for instance, it is not. To say that the intersection of two regular open sets is regular may sound plausible (this is what is involved in (3)), and it is true. It is, however, just as plausible to say that the union of two regular open sets is regular, but that is false. Example: let X be a plane and let P and Q be the open right half-plane and left half-plane. In intuitive terms, an open set is regular if there are no cracks in it; the trouble with the union of two regular open sets is that there might be a crack between them. This example helps to explain the necessity for the possibly surprising definition (4). It is obvious that something unusual, such as (5) for instance, is needed in the definition of complementation; the set-theoretic complement of an open set (regular or not) is quite unlikely to be open.

LEMMA 1. *If* $P \subset Q$, *then* $Q^{\perp} \subset P^{\perp}$.

Proof. Closure preserves inclusions and complementation reverses them.

LEMMA 2. *If* P *is open, then* $P \subset P^{\perp\perp}$.

Proof. Since $P \subset P^{-}$, it follows, by complementation, that $P^{\perp} \subset P'$. Now apply closure: since P' is closed, it follows that $P^{\perp-} \subset P'$, and this is the complemented version of what is wanted.

LEMMA 3. *If* P *is open, then* $P^{\perp} = P^{\perp\perp\perp}$.

Proof. Apply Lemma 1 to the conclusion of Lemma 2 to get $P^{\perp\perp\perp} \subset P^{\perp}$, and apply Lemma 2 to the open set $P^{\perp}$ (in place of P) to get the reverse inclusion.

It is an immediate consequence of Lemma 3 that if P is open, and all the more if P is regular, then $P^{\perp}$ is regular;

this proves that the right side of (5) belongs to the class A of regular open sets. Since $(P \cup Q)^{\perp}$ is always open, the same thing is true for (4). To settle (3), one more argument is needed.

LEMMA 4. *If P and Q are open, then* $(P \cap Q)^{\perp\perp} = P^{\perp\perp} \cap Q^{\perp\perp}$.

Proof. Since $P \cap Q$ is included in both P and Q, it follows from Lemma 2 that $P \cap Q$ is included in both $P^{\perp\perp}$ and $Q^{\perp\perp}$ and hence in their intersection. The reverse inclusion depends on the general topological fact that if P is open, then

$$P \cap Q^- \subset (P \cap Q)^-.$$

(If U is a neighborhood of a point of $P \cap Q^-$, then so is $U \cap P$, and this implies that $U \cap P$ meets Q, or equivalently, that U meets $P \cap Q$.) Complementing this relation we get

$$(P \cap Q)^{\perp} \subset P' \cup Q^{\perp}.$$

If now we apply closure and then complementation, then, since closure distributes over union and since P' is closed, it follows that

(6) $$P \cap Q^{\perp\perp} \subset (P \cap Q)^{\perp\perp}.$$

An application of (6) with $P^{\perp\perp}$ in place of P, followed by an application of (6) with the roles of P and Q interchanged, yields, via Lemma 1,

$$P^{\perp\perp} \cap Q^{\perp\perp} \subset (P^{\perp\perp} \cap Q)^{\perp\perp} \subset (P \cap Q)^{\perp\perp\perp\perp};$$

the conclusion follows from Lemma 3.

Lemma 4 implies immediately that the intersection of two regular open sets is regular, and hence that the right side of (3) belongs to A.

To complete the proof of Theorem 1 we must now show that the Boolean operations defined by (1)-(5) satisfy some system of axioms for Boolean algebras. It is less trouble to verify every one of the conditions (2.6)-(2.15) than to prove that some small subset of them is sufficient to imply the rest. In the verifications of (2.6), (2.7), (2.8), (2.10), (2.11), (2.12), (2.13), and (2.14) nothing is needed beyond the definitions, the equation $(P \cup Q)^{\perp} = P^{\perp} \cap Q^{\perp}$ (valid for any two sets P and Q), and trivial computations. The proof of (2.15) depends on Lemma 4. The fact (2.9) that $P \cap P^{\perp} = \emptyset$ is obvious (since $P^{\perp} \subset P'$). All that remains is to verify that $(P \cup P^{\perp})^{\perp\perp} = X$. This is not obvious; one way to go is by way of a little topological lemma that has other applications also.

LEMMA 5. *The boundary of an open set is a nowhere dense closed set.*

Proof. If P is open, and if the boundary of P included a non-empty open set, then that open set would have a non-empty intersection (namely itself) with P^{-}, and, at the same time, it would be disjoint from P. This contradicts the fundamental property of closure (often used as the definition).

Lemma 5 implies that if P is open, and all the more if it is regular, then the complement of the boundary of P, that is, $P \cup P^{\perp}$, is a dense open set. It follows that $(P \cup P^{\perp})^{\perp} = \emptyset$ and hence that $(P \cup P^{\perp})^{\perp\perp} = X$. This verifies (2.9) and completes the proof of Theorem 1.

Exercises

(1) If P and Q are open (and if $^\circ$ denotes the formation of interiors), then $(P \cap Q)^{-\circ} = (P^- \cap Q^-)^\circ$.

(2) If P is an arbitrary subset of a topological space, then $P^{\perp -} = P^{\perp \perp \perp -}$.

(3) What is the largest number of distinct sets obtainable from a subset of a topological space by repeated applications of closure and complementation? Construct an example for which this largest number is attained.

(4) Is the class of regular open sets always a base?

§ 5. Elementary relations

The least profound among the properties of an algebraic system are usually the relations among its elements (as opposed to the relations among subsets of it and functions on it). In this section we shall discuss some of the elementary relations that hold in Boolean algebras. Since we shall later meet a powerful tool (namely, the representation theorem for Boolean algebras) the use of which reduces all elementary relations to set-theoretic trivialities, the purpose of the present discussion is more to illustrate than to exhaust the subject. An incidental purpose is to establish some notation that will be used freely throughout the sequel.

Throughout this section p, q, r, $\cdots$ are elements of an arbitrary but fixed Boolean algebra A.

LEMMA 1. *If* $p \vee q = p$ *for all* p, *then* $q = 0$; *if* $p \wedge q = p$ *for all* p, *then* $q = 1$.

Proof. To prove the first assertion, put $p = 0$ and use (2.8); the second assertion is the dual of the first.

LEMMA 2. *If* p *and* q *are such that* $p \wedge q = 0$ *and* $p \vee q = 1$, *then* $q = p'$.

Proof. $q = 1 \wedge q = (p \vee p') \wedge q = (p \wedge q) \vee (p' \wedge q) = 0 \vee (p' \wedge q) = (p' \wedge p) \vee (p' \wedge q) = p' \wedge (p \vee q) = p' \wedge 1 = p'$.

These two lemmas can be expressed by saying that (2.8) uniquely determines 0 and 1, and (2.9) uniquely determines p'. In a less precise but more natural phrasing we may simply say that 0 and 1 and complementation are unique.

LEMMA 3. *For all* p *and* q, $p \vee (p \wedge q) = p$ *and* $p \wedge (p \vee q) = p$.

Proof. $p \vee (p \wedge q) = (p \wedge 1) \vee (p \wedge q) = p \wedge (1 \vee q) = p \wedge 1 = p$; the second equation is the dual of the first.

The identities of Lemma 3 are called the laws of *absorption*.

Often the most concise and intuitive way to state an elementary property of Boolean algebras is to introduce a new operation. Thus for instance, set-theoretic considerations suggest the operation of *subtraction*. We write

$$p - q = p \wedge q'.$$

The "symmetrized" version of the *difference* $p - q$ is the

Boolean sum:

$$(p - q) \vee (q - p) = p + q.$$

As a sample of the sort of easily proved relation that the notation suggests consider the distributive law

$$p \wedge (q - r) = (p \wedge q) - (p \wedge r).$$

One reason why Boolean algebras have something to do with logic is that the familiar sentential connectives *and*, *or*, and *not* have properties similar to the Boolean connectives $\wedge$, $\vee$, and $'$. Instead of meet, join, and complement, the logical terminology uses *conjunction*, *disjunction*, and *negation*. Motivated by the analogy, we now introduce into the study of Boolean algebra the operations suggested by logical *implication*,

$$p \Rightarrow q = p' \vee q,$$

and *biconditional*,

$$p \Leftrightarrow q = (p \Rightarrow q) \wedge (q \Rightarrow p).$$

The source of these operations suggests an unintelligent error that it is important to avoid. The result of the operation $\Rightarrow$ on the elements p and q of the Boolean algebra A is another element of A; it is not an assertion about or a relation between the given elements p and q. (The same is true of $\Leftrightarrow$.) It is for this reason that logicians sometimes warn against reading $p \Rightarrow q$ as "p implies q" and suggest instead the reading "if p, then q". Observe incidentally that if $\vee$ is read as "or", the disjunction $p \vee q$ must be interpreted in the non-exclusive sense (either p, or q, or both). The exclusive "or" (either p, or q, but not both) corresponds to the Boolean sum $p + q$.

The operations $\Rightarrow$ and $\Leftrightarrow$ would arise in any systematic study of Boolean algebra even without any motivation from logic. The reason is duality: the dual of $p - q$ is $q \Rightarrow p$, and the dual of $p + q$ is $p \Leftrightarrow q$. The next well-known Boolean operation that deserves mention here could not have been discovered through considerations of duality alone. It is called the (Sheffer) *stroke*, and it is defined by

$$p|q = p' \wedge q'.$$

In logical contexts this operation is known as *binary rejection* (neither p nor q).

The chief theoretical application of the Sheffer stroke is the remark that a single operation, namely the stroke, is enough to define Boolean algebras. To establish this remark, it is sufficient to show that complement, meet, and join can be expressed in terms of the stroke, and, indeed

$$p' = p|p,$$

$$p \wedge q = (p|p) \mid (q|q),$$

$$p \vee q = (p|q) \mid (p|q).$$

Exercises

(1) Prove that the following identities constitute a set of axioms for Boolean algebras:

$$(p|p) \mid (p|p) = p,$$

$$(p \mid (q|(q|q))) = p|p,$$

$$(p|(q|r)) \mid (p|(q|r)) = ((q|q)|p) \mid ((r|r)|p).$$

(2) Enumerate all possible binary operations on 2 (that is, mappings from 2×2 into 2). Identify each of these 16 operations in terms of operations introduced before.

(3) Show that if a ternary Boolean operation g is defined by

$$g(p, q, r) = (p \wedge q) \vee (q \wedge r) \vee (r \wedge p),$$

then that operation is enough to define Boolean algebras. Exhibit a set of axioms stated in terms of g, (Note that if g is regarded as defining a binary operation for each q, by, say,

$$p(q)r = g(p, q, r),$$

then $p(0)r = p \wedge r$ and $p(1)r = p \vee r$.)

§ 6. Order

We continue to work with an arbitrary but fixed Boolean algebra A.

LEMMA 1. *$p \wedge q = p$ if and only if $p \vee q = q$.*

Proof. If $p \wedge q = p$, then $p \vee q = (p \wedge q) \vee q$, and the conclusion follows from the appropriate law of absorption. The converse implication is obtained from this one by interchanging the roles of p and q and forming duals.

In set theory the corresponding equations characterize inclusion; that is, either one of the conditions $P \cap Q = P$ and $P \cup Q = Q$ is equivalent to $P \subset Q$. This motivates the introduction of a binary relation $\leqq$ in every Boolean algebra; we write

$$p \leqq q \text{ or } q \geqq p$$

in case $p \wedge q = p$, or, equivalently, $p \vee q = q$.

LEMMA 2. *The relation $\leqq$ is a partial order. In other words, it is reflexive ($p \leqq p$), antisymmetric (if $p \leqq q$, and*

q ≦ p, then p = q), and transitive (if p ≦ q and q ≦ r, then
p ≦ r).

Proof. The three conclusions follow, respectively, from the facts that ∧ and ∨ are idempotent (2.11), commutative (2.13), and associative (2.15).

It is sound mathematical practice to re-examine every part of a structure in the light of each new feature soon after the novelty is introduced. Here is the result of an examination of the structure of a Boolean algebra in the light of the properties of order.

LEMMA 3. (1) $0 \leqq p$ and $p \leqq 1$.

(2) *If $p \leqq q$ and $r \leqq s$, then $p \wedge r \leqq q \wedge s$ and $p \vee r \leqq q \vee s$.*

(3) *If $p \leqq q$, then $q' \leqq p'$.*

(4) *$p \leqq q$ if and only if $p - q = 0$, or, equivalently, $p \Rightarrow q = 1$.*

The proofs of all these assertions are automatic. It is equally automatic to discover the dual of ≦; according to any reasonable interpretation of the phrase it is ≧.

If E is any subset of a partially ordered set such as our Boolean algebra A, we can consider the set F of all upper bounds of E and ask whether or not F has a smallest element. In other words: an element q belongs to F in case $p \leqq q$ for every p in E; to say that F has a smallest element means that there exists an (obviously unique) element q_0 in F such that $q_0 \leqq q$ for every q in F. We shall call the least upper bound of the set E (if it has one) the *supremum* of E. All these considerations have their obvious duals. The greatest lower bound of E is called the *infimum* of E.

If the set E is empty, then every element of A is an upper bound of E (p in E implies $p \leq q$ for each q), and, consequently, E has a supremum, namely 0. Similarly (dually) if E is empty, then E has an infimum, namely 1.

Consider next the case of a singleton, say $\{p\}$. Since p itself is an upper bound of this set, it follows that the set has a supremum, namely p, and, similarly, that it has an infimum, namely p again.

The situation becomes less trivial when we pass to sets of two elements.

LEMMA 4. *For each p and q, the set $\{p, q\}$ has the supremum $p \vee q$ and the infimum $p \wedge q$.*

Proof. Since both p and q are dominated by $p \vee q$, that element is one of the upper bounds of $\{p, q\}$. It remains to prove that $p \vee q$ is the least upper bound, or, in other words, that if both p and q are dominated by some element r, then $p \vee q \leq r$. This is easy; by (2), $p \vee q \leq r \vee r$. The assertion about infimum follows by duality.

Lemma 4 generalizes immediately to arbitrary non-empty finite sets (instead of sets with only two elements). We may therefore conclude that if E is a non-empty finite subset of A, then E has both a supremum and an infimum, namely $\vee E$ and $\wedge E$, respectively. Motivated by these facts we hereby extend the interpretation of the symbols used for joins and meets to sets that may be empty or infinite. If a subset E of A has a supremum, we shall denote that supremum by $\vee E$ regardless of the size of E, and, similarly we shall use $\wedge E$ for all infima. In this notation what we know about very small sets can be expressed as follows: $\vee \emptyset = 0$, $\wedge \emptyset = 1$, and $\vee \{p\} = \wedge \{p\} = p$. The

notation used earlier for the join or meet of a finite sequence of elements is also extendable to the infinite case.
Thus if $\{p_i\}$ is an infinite sequence with a supremum
(properly speaking: if the range of the sequence has a
supremum), then that supremum is denoted by $\bigvee_{i=1}^{\infty} p_i$. If,
more generally, $\{p_i\}$ is an arbitrary family with a supremum,
indexed by the elements i of a set I, the supremum is
denoted by $\bigvee_{i \in I} p_i$, or, in case no confusion is possible,
simply by $\bigvee_i p_i$.

Exercises

(1) The concept of divisibility makes sense in every
ring: p is divisible by q in case $p = qr$ for some r. In a
Boolean ring, p is divisible by q if and only if $p \leqq q$.

(2) True or false: if $p \leqq q$ and $r \leqq s$, then $p + r \leqq q + s$
and $p \Longleftrightarrow r \leqq q \Longleftrightarrow s$?

(3) A *lattice* is a partially ordered set ın which every
set of two elements has both a supremum and an infimum.
In analogy with Boolean algebras, the supremum and infimum
of $\{p, q\}$ are denoted by $p \vee q$ and $p \wedge q$, respectively. Prove
that, in every lattice, the identities (the distributive laws)
$p \wedge (q \vee r) = (p \wedge q) \vee (p \wedge r)$ and $p \vee (q \wedge r) = (p \vee q) \wedge (p \vee r)$
imply each other. A lattice in which they hold is called
distributive.

(4) A lattice is called *complemented* if it contains two
elements 0 and 1 such that $0 \leqq p$ and $p \leqq 1$ for all p, and such
that, corresponding to each p, there exists at least one q
with $p \wedge q = 0$ and $p \vee q = 1$. Prove that in a distributive
lattice complementation is unique.

(5) Interpret and prove the assertion: a complemented distributive lattice is a Boolean algebra.

§ 7. Infinite operations

An infinite subset of a Boolean algebra may fail to have a supremum. (Example: take the finite-cofinite algebra of integers and consider the singletons of the even integers.) A Boolean algebra with the property that every subset of it has both a supremum and an infimum is called a *complete* (Boolean) *algebra*. Similarly, a field of sets with the property that both the union and the intersection of every class of sets in the field is again in the field is called a *complete field* of sets. The simplest example of a complete field of sets (and hence of a complete algebra) is the field of all subsets of a set. Our next example of a complete algebra is not a field; it is the regular open algebra of a topological space (cf. Theorem 1, p. 13). For purposes of reference it is worth while recording the formal statement.

LEMMA 1. *The regular open algebra of a topological space is a complete Boolean algebra. The supremum and the infimum of a family* $\{P_i\}$ *of regular open sets are* $(\bigcup_i P_i)^{\perp\perp}$ *and* $(\bigcap_i P_i)^{\perp\perp}$ *, respectively.*

Proof. If $(\bigcup_i P_i)^{\perp\perp} = P$, then since each P_i is included in their union, Lemma 4.2 implies that $P_i \subset P$ for every i. (Since the meet of two regular open sets is the same as their intersection, it follows that the Boolean order relation for regular open sets is the same as ordinary set-theoretic inclusion.) To prove that the upper bound P is the least possible one, suppose that Q is a regular open set such that $P_i \subset Q$ for every i. The proof that then $P \subset Q$ is quite easy: just observe that $\bigcup_i P_i \subset Q$ and apply Lemma 4.1

twice to obtain $P \subset Q^{\perp\perp} = Q$. The characterization of infima proceeds dually.

The last sentence of the preceding proof is justifiable, but, perhaps, a trifle premature. It leans implicitly on the following infinite versions of the DeMorgan laws.

LEMMA 2. *If $\{p_i\}$ is a family of elements in a Boolean algebra, then*

$$(\bigvee_i p_i)' = \bigwedge_i p_i' \quad and \quad (\bigwedge_i p_i)' = \bigvee_i p_i'.$$

The equations are to be interpreted in the sense that if either term in either equation exists, then so does the other term of that equation, and the two terms are equal.

Proof. Suppose $p = \bigvee_i p_i$. Since $p_i \leq p$ for every i, it follows that $p' \leq p_i'$ for every i. It is to be proved that if $q \leq p_i'$ for every i, then $q \leq p'$. The assumption implies that $p_i \leq q'$ for every i, and hence, from the definition of supremum, $p \leq q'$. A dual argument justifies the passage from the left side of the second equation to the right. To justify the reverse passage, apply the results already proved to the families of complements.

COROLLARY. *If every subset of a Boolean algebra has a supremum (or else if every subset has an infimum), then that algebra is complete.*

It will usually not be sufficient to know merely that certain infinite suprema exist; the algebraic properties of those suprema (such as commutativity, associativity, and distributivity) are also needed.

It is almost meaningless to speak of infinite commutative laws. An infinite supremum is something associated with a set of elements, and, by definition, it is independent of any possible ordering of that set.

A reasonable verbal formulation of an infinite associative law might go like this. Form each of several suprema and then form their supremum; the result should be equal to the supremum of all the elements that originally contributed to each separate supremum. It is worth while to state and prove this in a more easily quotable form.

LEMMA 3. *If $\{I_j\}$ is a disjoint family of sets with union I, and if p_i, for each i in I, is an element of a Boolean algebra, then*

$$\bigvee{}_j \left(\bigvee{}_{i \, \epsilon \, I_j} p_i \right) = \bigvee{}_{i \, \epsilon \, I} p_i.$$

The equation is to be interpreted in the sense that if the left side exists, then so does the right, and the two are equal.

Proof. Write $q_j = \bigvee_{i \, \epsilon \, I_j} p_i$ and $q = \bigvee_j q_j$. We are to prove that q is an upper bound of the family $\{p_i : i \, \epsilon \, I\}$, and that, in fact, it is the least upper bound. Since each i in I belongs to exactly one I_j, it follows that for each i there is one j with $p_i \leqq q_j$; since, moreover, $q_j \leqq q$, it follows that q is indeed an upper bound. Suppose now that $p_i \leqq r$ for every i. Since, in particular, $p_i \leqq r$ for every i in I_j, it follows from the definition of supremum that $q_j \leqq r$. Since this is true for every j, we may conclude, similarly, that $q \leqq r$, and this completes the proof.

The preceding comments on infinite commutativity and associativity were made for suprema; it should go without saying that the corresponding (dual) comments for infima are just as true. The most interesting infinite laws are the ones in which suprema and infima occur simultaneously. These are the distributive laws, to which we now turn. They too come in dual pairs; we shall take advantage of the principle of duality and restrict our attention to only one member of each such pair. We begin with the simplest infinite distributive law.

LEMMA 4. $p \wedge \bigvee_i q_i = \bigvee_i (p \wedge q_i)$. *The equation is to be interpreted in the sense that if the left side exists, then so does the right, and the two are equal.*

Proof. Write $q = \bigvee_i q_i$; clearly $p \wedge q_i \leq p \wedge q$ for every i. It is to be proved that if $p \wedge q_i \leq r$ for every i, then $p \wedge q \leq r$. For the proof, observe that

$$q_i = (p \wedge q_i) \vee (p' \wedge q_i) \leq r \vee p',$$

and hence, by the definition of supremum,

$$q \leq r \vee p'.$$

Form the meet of both sides of this inequality with p to get

$$p \wedge q \leq p \wedge r;$$

the desired conclusion now follows from the trivial fact that $p \wedge r \leq r$.

To motivate the most restrictive distributive law, consider a long infimum of long suprema, such as

$$(p_{11} \vee p_{12} \vee p_{13} \vee \cdots) \wedge (p_{21} \vee p_{22} \vee p_{23} \vee \cdots) \wedge (p_{31} \vee p_{32} \vee p_{33} \vee \cdots) \wedge \cdots .$$

Algebraic experience suggests that this ought to be equal to a very long supremum, each of whose terms is a long infimum like $p_{12} \wedge p_{23} \wedge p_{31} \wedge \cdots$. The way to get all possible infima of this kind is to pick one term from each original supremum in all possible ways. The picking is done, of course, by a function that associates with each value of the first index some value of the second index; the "very long" supremum has one term corresponding to each such function.

We are now ready for a formal definition. Suppose that I and J are two index sets and that $p(i, j)$ is an element of a Boolean algebra A whenever $i \in I$ and $j \in J$. Let J^I be the set of all functions from I to J. We say that the family $\{p(i, j)\}$ satisfies the *complete distributive law* in case

(1) $\quad \bigwedge_{i \in I} \bigvee_{j \in J} p(i, j) = \bigvee_{a \in J^I} \bigwedge_{i \in I} p(i, a(i)).$

The assertion of the equation is intended here to imply, in particular, the existence of the suprema and infima that occur in it. If the algebra A is such that the existence of either side of (1) (for every family $\{p(i, j)\}$) implies that the other side also exists and that the two are equal, then A is called *completely distributive*.

The field of all subsets of a set is always completely distributive. The regular open algebra of a topological space may fail to be so. Consider, for instance, the regular open algebra of the open unit interval $(0, 1)$. (Warning to the would-be expert. Compactness, or its absence, has nothing to do with this example; the endpoints were omitted for notational convenience only.) Let I be the set of positive integers and let J be the set consisting of the two numbers $+1$ and -1. To define $P(i, j)$, cut up the interval into 2^i open intervals of length 2^{-i}; let $P(i, +1)$ be the union of the open left halves of these intervals and let $P(i, -1)$ be the union of their open right halves. Since $P(i, +1) \vee P(i, -1)$ is equal to the entire space $(0, 1)$ for each i, it follows that the left side of (1) is the unit element of the algebra under consideration. A moment's reflection on the binary expansions of real numbers shows that

$\bigcap_i P(i, \alpha(i))$ consists of at most one point, whatever the function α in J^I may be; it follows that the right side of (1) is the zero element of our algebra.

Exercises

(1) Is a complete field of subsets of a set X the same as the field of all subsets of X?

(2) Give an example of a field of sets that happens to be a complete Boolean algebra but not a complete field of sets.

(3) It follows from Lemmas 1 and 2 that if $\{P_i\}$ is a family of regular open sets, then

$$(\bigcap_i P_i)^{-\prime-\prime} = (\bigcap_i P_i^-)^{\prime-\prime}.$$

Show that this is not necessarily true for arbitrary open sets and give a direct topological proof for regular open sets.

(4) If a Boolean algebra is such that every subset of it has either a supremum or an infimum, is it necessarily complete?

(5) Interpret and prove the equation

$$p \vee \bigvee_i p_i = \bigvee_i (p \vee p_i).$$

(6) Interpret and prove the assertion: if for every i there is a j such that $p_i \leq q_j$, then

$$\bigvee_i p_i \leq \bigvee_j q_j.$$

(7) Interpret and prove the assertion: if $I \subset J$, then

$$\bigvee_{i \in I} p_i \leqq \bigvee_{i \in J} p_i \cdot$$

(8) Interpret and prove the equation

$$\bigvee_i p_i \wedge \bigvee_j q_j = \bigvee_{i,j} (p_i \wedge q_j).$$

(9) Discuss possible interpretations of the equations in Lemma 3 and Lemma 4 besides the ones there stated.

§ 8. Subalgebras

A Boolean *subalgebra* of a Boolean algebra A is a subset B of A such that B, together with the distinguished elements and operations of A, is a Boolean algebra.

Warning: the distinguished elements 0 and 1 are essential parts of the structure of a Boolean algebra. A subring of a ring with unit may or may not have a unit, and, if it has one, its unit may or may not be the same as the unit of the whole ring. For Boolean algebras this indeterminacy is defined away: a subalgebra must contain the element 1. The insistence on the role of 1 is not an arbitrary convention, but a theorem. Since complementation is indubitably an essential part of the structure of a Boolean algebra, the presence of 1 in every subalgebra can be proved. Proof: a subalgebra contains, along with each element p, the complement p' and the join $p \vee p'$. This proof made implicit use of the fact that a subalgebra is not empty. If 0 and 1 are not built into the definition of a Boolean subalgebra, then non-emptiness must be explicitly assumed.

To illustrate the situation, let Y be a non-empty subset of a set X. Both $\mathcal{P}(X)$ and $\mathcal{P}(Y)$ are Boolean algebras in a natural way (§2), and clearly every element of $\mathcal{P}(Y)$ is an element of $\mathcal{P}(X)$. Since, however, the unit of $\mathcal{P}(X)$ is X, whereas the unit of $\mathcal{P}(Y)$ is Y, it is not true that $\mathcal{P}(Y)$ is a Boolean subalgebra of $\mathcal{P}(X)$. Another reason why it is not true is, of course, that complementation in $\mathcal{P}(Y)$ is not the restriction of complementation in $\mathcal{P}(X)$.

There is another possible source of misunderstanding, but one that is less likely to lead to error. (Reason: it is not special to Boolean algebras, but has its analogue in almost every algebraic system.) To be a Boolean subalgebra it is not enough to be a subset that is a Boolean algebra in its own right, however natural the Boolean operations may appear. The Boolean operations of a subalgebra, by definition, must be the restrictions of the Boolean operations of the whole algebra. The situation is illuminated by the regular open algebra A of a topological space X (§4). Clearly A is a subclass of the field $\mathcal{P}(X)$, but, equally clearly, A is not a subalgebra of $\mathcal{P}(X)$.

Every Boolean algebra A includes a *trivial* subalgebra, namely 2; all other subalgebras of A will be called *nontrivial*. Every Boolean algebra A includes an *improper* subalgebra, namely A; all other subalgebras will be called *proper*.

The definition of a field of subsets of a set X may be formulated by saying that it is a Boolean subalgebra of the special field $\mathcal{P}(X)$. In general a Boolean subalgebra of a field of sets is called a *subfield*. Here are two examples of subalgebras (in fact subfields): the finite-cofinite algebra of a set X is a subalgebra of the countable-cocountable algebra of X, and the Borel algebra of the real line is a subalgebra of the Lebesgue algebra.

If a non-empty subset B of a Boolean algebra A is closed under some Boolean operations, and if there are enough of those operations that all other Boolean operations can be defined by them, then B is a subalgebra of A. Example: if B is closed under joins and complements, then B is a subalgebra; alternatively, if B is closed under the Sheffer stroke, then B is a subalgebra.

A moment's thought shows that the intersection of every collection of subalgebras of a Boolean algebra A is again a subalgebra of A. It follows that if E is an arbitrary subset of A, then the intersection of all those subalgebras that happen to include E is a subalgebra. (There is always at least one subalgebra that includes E, namely the improper subalgebra A.) That intersection, say B, is the smallest subalgebra of A that includes E; in other words, B is included in every subalgebra that includes E. The subalgebra B is called the subalgebra *generated* by E. Thus, for example, if E is empty, then the subalgebra generated by E is the smallest possible subalgebra of A, namely 2. A generating subset E of a subalgebra B is also known as a set of *generators* of B.

The definition of a Boolean subalgebra B says nothing about the infinite suprema and infima that may be formable in the whole algebra A. Anything can happen: suprema or infima can be gained or lost or change value as we pass back and forth between A and B. Everything that can happen can be illustrated in the theory of complete Boolean algebras. If B is a subalgebra of a complete algebra A, and if the supremum (in A) of every subset of B belongs to B, we say that B is a *complete subalgebra* of A. (Warning: this is stronger than requiring merely that B be a complete Boolean algebra in its own right.) Note that a complete subalgebra of A contains the infima (in A) of all its subsets, as well as their suprema. In the case of fields we speak of *complete subfields*. For complete algebras the concept of a generated complete subalgebra is defined the same way as when completeness was not yet mentioned; all that is necessary is to replace "algebra" by "complete algebra" throughout the discussion.

Exercises

(1) A subring of a Boolean ring need not be a Boolean subalgebra; what if the subring contains 1?

(2) Every subset of a partially ordered set inherits a partial order from the whole·set. If a non-empty subset of a Boolean algebra is construed as a partially ordered set in this way, and if it turns out that with respect to this partial order it is a complemented distributive lattice, does it follow that it is a Boolean subalgebra of the original algebra? (See Exercise 6.5.)

(3) If a subset B of a Boolean algebra A contains 0 and 1 and is closed under the formation of meets and joins, does it follow that B is a subalgebra of A?

(4) Give an example of a Boolean subalgebra B of a Boolean algebra A and of a subset E of B such that E has a supremum in B but not in A.

(5) Prove that an infinite Boolean algebra with m generators has m elements.

(6) Suppose that a subalgebra B of a Boolean algebra A is such that whenever a subset E of B has a supremum p in B, then p is the supremum of E in A also. A subalgebra satisfying this condition is sometimes called *regular*. Prove

that a necessary and sufficient condition that B be a regular subalgebra of A is that whenever E is a subset of B with $\wedge E = 0$ in B, then $\quad \wedge E = 0$ in A also.

(7) Is a complete subalgebra of a complete algebra necessarily regular? Is a regular subalgebra of a regular subalgebra a regular subalgebra?

§ 9. Homomorphisms

A *Boolean homomorphism* is a mapping f from a Boolean algebra B, say, to a Boolean algebra A, such that

(1) $$f(p \wedge q) = f(p) \wedge f(q),$$

(2) $$f(p \vee q) = f(p) \vee f(q),$$

(3) $$f(p') = (f(p))',$$

whenever p and q are in B. In a somewhat loose but brief and suggestive phrase, a homomorphism is a structure-preserving mapping between Boolean algebras. A convenient synonym for "homomorphism from B to A" is "A-valued homomorphism on B". Such expressions will be used most frequently in case $A = 2$.

Special kinds of Boolean homomorphisms may be described in the same words as are used elsewhere in algebra. A homomorphism may be one-to-one into (monomorphism, if $f(p) = f(q)$, then $p = q$); it may be onto (epimorphism, every element of A is equal to $f(p)$ for some p in B); it may be both one-to-one and onto (isomorphism); its range may be included in its domain (endomorphism, $A \subset B$); and it may be a one-to-one mapping of its domain onto itself (automorphism). If there exists an isomorphism from B onto A, then A and B are called *isomorphic*.

The distinguished elements 0 and 1 play a special role for homomorphisms, just as they do for subalgebras. Indeed, if f is a Boolean homomorphism and p is an element in its domain ($p = 0$ will do), then

$$f(p \wedge p') = f(p) \wedge (f(p))',$$

and, therefore,

(4) $$f(0) = 0.$$

This much would be expected by a student of ring theory. What is important is that the dual argument proves the dual fact,

(5) $$f(1) = 1.$$

The mapping that sends every element of one Boolean algebra onto the zero element of another is simply not a homomorphism; in the theory of Boolean algebras there is no such thing as a "trivial" homomorphism.

The equations (1) and (2) imply that 0 and 1 belong to the range of every homomorphism; a glance at the equations (1)-(3) should complete the proof that the range of every homomorphism, from B into A say, is a Boolean subalgebra of A. The range of a homomorphism with domain B is called a *homomorphic image* of B.

Since every Boolean operation (e.g., + and $\implies$) can be defined in terms of $\wedge$, $\vee$, and $'$, it follows that a Boolean homomorphism preserves all such operations. If, that is, f is a Boolean homomorphism and p and q are elements of its domain, then

$$f(p + q) = f(p) + f(q) \text{ and } f(p \Rightarrow q) = f(p) \Rightarrow f(q).$$

It follows, in particular, that every Boolean homomorphism is a ring homomorphism, and also that every Boolean homomorphism is order-preserving. The last assertion means that if $p \leqq q$, then $f(p) \leqq f(q)$.

The crucial fact in the preceding paragraph was the definability of Boolean operations and relations in terms of meet, join, and complement. Thus, more generally, if a mapping f from a Boolean algebra B to a Boolean algebra A preserves enough Boolean operations so that all others are definable in terms of them, then f is a homomorphism. Example: if f preserves $\vee$ and $'$ (that is, satisfies the identities (2) and (3)), then f is a homomorphism; alternatively, if f preserves the Sheffer stroke, then f is a homomorphism.

We proceed to consider some examples of Boolean homomorphisms.

For our first example let B be an arbitrary Boolean algebra, and let p_0 be an arbitrary non-zero element of B. The set A of all subelements of p_0 (this means the elements p with $p \leqq p_0$) can be construed as a Boolean algebra, as follows: 0, meet, and join in A are the same as in B, but 1 and p' in A are defined to be the elements p_0 and $p_0 - p$ of B. The mapping $p \rightarrow p \wedge p_0$ is an A-valued homomorphism on B.

Consider next a field B of subsets of a set X, and let x_0 be an arbitrary point of X. For each set P in B, let $f(P)$ be 1 or 0 according as $x_0 \in P$ or $x_0 \in' P$. The mapping f is a

2-valued homomorphism on B. Observe that $f(P)$ is equal to the value of the characteristic function of P at x_0.

For one more example, let ϕ be an arbitrary mapping from a non-empty set X into a set Y, and let A and B be fields of subsets of X and Y respectively. Write $f = \phi^{-1}$, or, explicitly, for each P in B, let $f(P)$ be the inverse image of P. In general the set $f(P)$ will not belong to the field A. If $f(P) \in A$ whenever $P \in B$, then f is an A-valued homomorphism on B.

For purposes of reference we shall call the homomorphisms described in these three examples the homomorphisms *induced* by p_0, x_0, and ϕ, respectively.

If B is a subalgebra of an algebra A, then the identity mapping (that is, the mapping f defined for every p in B by $f(p) = p$) is a homomorphism from B into A, and, in particular, the identity mapping on A is an automorphism of A. There is a natural way to define the product of (some) pairs of homomorphisms, and it turns out that the identity mappings just mentioned indeed act as multiplicative identities. The *product* (or *composite*) $f \circ g$ of two homomorphisms f and g is defined in case A, B, and C are Boolean algebras, f maps B into A, and g maps C into B; the value of $f \circ g$ at each element p of C is given by

$$(f \circ g)\,(p) = f(g(p)).$$

If, moreover, h is a homomorphism from D, say, to C, then

$$f \circ (g \circ h) = (f \circ g) \circ h,$$

that is, the operation of composition is associative.

An isomorphism between Boolean algebras preserves every infinite supremum and infimum that happens to exist, but, in general, a mere homomorphism will not do so. A homomorphism f is called complete in case it preserves all suprema (and, consequently, all infima) that happen to exist. This means that if $\{p_i\}$ is a family of elements in the domain of f with supremum p, the family $\{f(p_i)\}$ has a supremum and that supremum is equal to $f(p)$.

Exercises

(1) Is every ring homomorphism between Boolean algebras a Boolean homomorphism? What if it preserves 1?

(2) If a mapping f between Boolean algebras preserves 0, 1, $\wedge$, and $\vee$, is it necessarily a Boolean homomorphism?

(3) If a mapping f between Boolean algebras preserves order, is it necessarily a Boolean homomorphism?

(4) Suppose that both f and g are A-valued homomorphisms on B. Define a mapping $f \vee g$ from B into A by

$$(f \vee g)(p) = f(p) \vee g(p).$$

Is $f \vee g$ a homomorphism? What about $f + g$ (defined similarly)?

(5) Prove that if E generates B, and if f and g are A-valued homomorphisms on B such that $f(p) = g(p)$ whenever $p \in E$, then $f = g$. What if B is the complete algebra generated by E and f and g are complete homomorphisms?

(6) Give an example of an incomplete homomorphism between complete Boolean algebras. Can such an example be a monomorphism? An epimorphism?

(7) Prove that if f is a Boolean homomorphism between complete algebras, and if $\{p_i\}$ is a family of elements in the domain of f, then

$$\bigvee_i f(p_i) \leqq f(\bigvee_i p_i).$$

(8) Prove that if a subalgebra B of a Boolean algebra A happens to be complete (considered as an algebra in its own right), then a necessary and sufficient condition that B be a complete subalgebra of A is that the identity mapping of B into A be a complete homomorphism.

(9) Prove that a subalgebra B of a Boolean algebra A is regular if and only if the identity mapping of B into A is a complete homomorphism.

(10) Prove that the range of a homomorphism is a regular subalgebra if and only if the homomorphism is a complete homomorphism.

§10. Free algebras

The elements of every subset of every Boolean algebra satisfy various algebraic conditions (such, for example, as the distributive laws) just by virtue of belonging to the same Boolean algebra. If the elements of some particular set E satisfy no conditions except these necessary universal ones, it is natural to describe E by some such word as "free." A crude but suggestive way to express the fact that the elements of E satisfy no special conditions is to say that the elements of E can be transferred to an arbitrary Boolean algebra in a completely arbitrary way with no danger of encountering a contradiction. In what follows we shall make these heuristic considerations precise. We shall restrict attention to sets that generate the entire algebra; from the practical point of view the loss of generality involved in doing so is negligible.

A set E of generators of a Boolean algebra B is called *free* if every mapping from E to an arbitrary Boolean algebra A can be extended to an A-valued homomorphism on B. In more detail: E is free in case for every Boolean algebra A and for every mapping g from E into A there exists an A-valued homomorphism f on B such that $f(p) = g(p)$ for every p in E. Equivalent expressions: "E freely generates B", or even "B is free on E". A Boolean algebra is called free if it has a free set of generators.

The definition is conveniently summarized by the subjoined diagram.

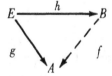

The diagram is to be interpreted as follows. The arrow h is the identity mapping from E to B, expressing the fact that E is a subset of B. The arrow g is an arbitrary mapping from E to an arbitrary algebra A. The arrow f is, of course, the homomorphic extension required by the definition; it is dotted to indicate that it comes last, as a construction based on h and g. It is understood that the diagram is "commutative" in the sense that $(f \circ h)(p) = g(p)$ for every p in E.

The arrow diagram does not express the fact that E generates B. The most useful way that that fact affects the mappings under consideration is to guarantee uniqueness: there can be only one A-valued homomorphism f on B that agrees with g on E. One way of expressing this latter fact is to say that f is uniquely determined by g and h.

There is another and even more important uniqueness assertion that can be made here. If B_1 and B_2 are Boolean algebras, free on subsets E_1 and E_2, respectively, and if E_1 and E_2 have the same cardinal number, then B_1 and B_2 are isomorphic, via an isomorphism that interchanges E_1 and E_2. This says, roughly speaking, that B is uniquely determined (to within isomorphism) by the cardinal number of E. The proof is summarized by the diagram:

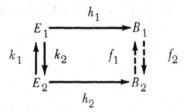

Here k_1 is a one-to-one mapping from E_1 onto E_2 with inverse k_2, h_1 and h_2 are the given embeddings, and f_1 and f_2 are the homomorphic extensions of $h_2 \circ k_1$ and $h_1 \circ k_2$, respectively. The commutativity of the diagram tells us that the two ways of going from E_1 to B_2 must coincide, and the same is true for the two ways of going from F_2 to B_1. If we apply the appropriate one of f_1 and f_2 to these equalities, and use the fact that the composite of k_1 and k_2, in either order, is the identity on its domain, we can conclude that $f_2 \circ f_1$ and $f_1 \circ f_2$ are extensions of h_1 and h_2, respectively. Since the identity homomorphisms on B_1 and B_2 are also such extensions, the already known uniqueness result guarantees that the composite of f_1 and f_2, in either order, is the identity on its domain. This implies that f_1 and f_2 are isomorphisms, and the proof is complete.

There is one big gap in what we have seen so far of the theory of freely generated algebras. We may know all about uniqueness, but we know nothing about existence. The main thing to be known here is that for each cardinal number there actually exists a Boolean algebra that is free on a set having exactly that many elements. A somewhat unpleasant combinatorial proof of this existence theorem is available to us now. We shall not enter into it; we choose, instead, a pleasanter and more economical road. We postpone the existence proof till after the introduction of some powerful techniques. The purpose of this section is just to state the problem and to indicate, in bare outline, the combinatorial approach to its solution. The main virtue of this combinatorial approach is that it shows how Boolean algebras (and, in particular, free Boolean algebras) arise in considerations of logic.

A general theory of the usual sentential connectives (conjunction, disjunction, negation, implication, etc.) should be applicable to every conceivable collection of sentences. This implies that its basic constituents (generators) should be as unrestricted (free) as possible. Suppose now that we want to construct a theory equipped to deal with, say, at least m sentences simultaneously, where m is a cardinal number. The thing to do then is to take a set E of cardinality m, and to consider all the formal expressions obtained by combining the elements of E and the sentential connectives in an intelligent manner. Ultimately the elements of E are to be replaced (or, at any rate, replaceable) by sentences. All this can be done, and, incidentally, it is important that in the doing of it the cardinal number m should be allowed to be infinite. Even if a mathematician or logician wishes to consider only finite combinations of sentences, it seems both practically and theoretically undesirable to place a fixed upper bound on the number of sentences that may be combined. The only way to make one theory elastic enough to deal with all finite combinations is to provide it with an infinite supply of things that it may combine.

To achieve the desired end a logician will usually begin by selecting enough sentential connectives so that all others are definable in terms of them; we know, for instance, that $\vee$ (or) and $'$ (not) will do. Next, given the set E, the logician will proceed to form all finite sequences whose terms are the selected connectives, or elements of E, or parentheses, put together in the usual and obvious manner. Precisely speaking, the admissible sequences consist of: the one-term sequences whose term belongs to E; the sequences obtained by inserting $\vee$ between two others already admitted and enclosing the result in parentheses; the sequences obtained by following an already admitted sequence by $'$ and enclosing the result in parentheses; and no others. The reason for the insistence on parentheses is caution. The distinction between $(p \vee (q'))$ and $((p \vee q)')$ is obvious, whereas the customary decision that $p \vee q'$ means the former and not the latter is the result of quite an arbitrary and frequently unformulated convention. One other word of supercaution deserves mention: it must be assumed that neither the selected connectives nor the parentheses that are used occur as elements of E.

If the sequences so obtained are to form a part of a general theory of sentences, it is clear that certain identifications will have to be made. The sequence $(p \vee q)$ is different from $(q \vee p)$, but, if p and q are sentences, then "p or q" and "q or p" are, in some sense, the same sentence. The customary way to specify the identifications that sound logical intuition and practice demand is first to define a special class of admissible sequences (called *tautologies*) and then to say that two admissible sequences are to be identified just in case a certain easily describable combination of them is a tautology. The procedure is similar to the formation of quotient groups: first we select a normal subgroup and then we say that two elements of the given

group are congruent modulo that subgroup just in case
their quotient belongs to the selected normal subgroup.

To define the set of tautologies we first define certain
quite natural abbreviations, then, using these, we describe
some tautologies, and finally we obtain all tautologies by
describing a simple operation that makes new tautologies
out of old. The abbreviations are these: if S and T are
admissible sequences, we write $S \wedge T$ for $((S') \vee (T'))'$,
we write $S \Rightarrow T$ for $(S') \vee T$, and we write $S \Leftrightarrow T$ for
$(S \Rightarrow T) \wedge (T \Rightarrow S)$. The initial set of tautologies consists
of all the sequences of one of the four forms

$$(S \vee S) \Rightarrow S,$$

$$S \Rightarrow (S \vee T),$$

$$(S \vee T) \Rightarrow (T \vee S),$$

$$(S \Rightarrow T) \Rightarrow ((R \Rightarrow S) \Rightarrow (R \Rightarrow T)),$$

where R, S, and T are admissible sequences. (Each sequence of each of these forms is called an *axiom*.) The way
to make new tautologies out of old is this: if S is a tautology and if $S \Rightarrow T$ is a tautology, then T is a tautology.
(This operation is a *rule of inference*, namely, in classical
terms, *modus ponens*.) A tautology is, by definition, a
sequence that is either an axiom or obtainable from the
axioms by a finite number of applications of modus ponens.

Two sequences S and T are called *logically equivalent* in
case $S \Leftrightarrow T$ is a tautology.

The structure outlined in this way, that is, the structure
consisting of the set of all admissible sequences, the subset of tautologies, and the relation of logical equivalence

is known as the *propositional calculus*. The connection
between the propositional calculus (based, as above, on a
set of power m, say) and the theory of Boolean algebras is
this: logical equivalence is an equivalence relation, the
set of equivalence classes has in a natural way the structure of a Boolean algebra, and, in fact, that Boolean
algebra is freely generated by m generators.

The involved construction of the propositional calculus
outlined above is similar to, but definitely not identical
with, a well-known construction of free groups (via "words"
and equivalence classes). That familiar construction could
also be adapted to the construction of free Boolean algebras;
the result would be about equally painful with what we have
already seen. It is unimportant but amusing to know that
the cross-fertilization between the two theories is complete:
the "axiom-rule" approach can be adapted to the construction of free groups.

The early part of the theory of free Boolean algebras
extends with no profound conceptual change to the category
of complete algebras. The definition reads just as before
except that all the Boolean algebras that enter into it, and
all the homomorphisms also, are now required to be complete.
The uniqueness theorems are proved just as before. The
situation of the principal existence theorem, however, is
startlingly different. Both H. Gaifman and A. W. Hales
(Cal. Tech. thesis, 1962) have proved that for each cardinal
number m there exists a countably generated complete
Boolean algebra with m or more elements. (Generation is to
be interpreted here in the sense appropriate to the category
of complete Boolean algebras.)

Exercises

(1) What is the Boolean algebra freely generated by the empty set?

(2) For which sets X is the algebra $\mathcal{P}(X)$ free? Examine especially sets X of cardinality 1, 2, 3, and $\aleph_0$.

(3) Prove that a necessary and sufficient condition that a set E of generators of a Boolean algebra B be free is that whenever $p_1, \cdots, p_n$ are elements of B such that, for each i, either p_i or p_i' belongs to E, then $\bigwedge_{i=1}^{n} p_i \neq 0$.

(4) If X is an infinite set, is $\mathcal{P}(X)$ free?

(5) Is every subalgebra of a free algebra free?

§11. Ideals and filters

If f is a Boolean homomorphism, from B to A say, the *kernel* of f is the set of those elements in B that f maps onto 0 in A. In symbols the kernel M of f is defined by

$$M = f^{-1}(\{0\}),$$

or, equivalently, by

$$M = \{p : f(p) = 0\}.$$

Motivated by the immediately obvious properties of kernels, we make the following definition: a *Boolean ideal* in a Boolean algebra B is a subset M of B such that

(1) $0 \in M$,

(2)   if $p \in M$ and $q \in M$, then $p \vee q \in M$,

(3)   if $p \in M$ and $q \in B$, then $p \wedge q \in M$.

Clearly the kernel of every Boolean homomorphism is a Boolean ideal. Observe that condition (1) in the definition can be replaced by the superficially less restrictive condition that M be not empty, without changing the concept of ideal. Indeed, if M is not empty, say $p \in M$, and if M satisfies (3), then $p \wedge 0 \in M$.

The concept of Boolean ideal can also be defined in either ring-theoretic or order-theoretic terms. In the language of ring theory it turns out that an ideal is an ideal, or, to put it more precisely, that a subset M of a Boolean algebra B is a Boolean ideal if and only if it is an ideal in the Boolean ring. Suppose, indeed, that M is a Boolean ideal; it is to be proved that if p and q are in M, then $p + q$ is in M. The proof is easy: $p \wedge q' \in M$ by (3), $p' \wedge q \in M$ for the same reason, and consequently, $(p \wedge q') \vee (p' \wedge q) \in M$ by (2) (see (5)). Now suppose, conversely, that M is an ideal in the sense of ring theory; it is to be proved that if p and q are in M, then $p \vee q$ is in M. The proof is, if anything, easier than before: an ideal in a ring always contains $p + q + pq$ along with p and q (see (2.2)). The language of order does not have much to contribute to ideal theory. This much can be said: the condition (3) can be replaced by

if $p \in M$ and $q \leqq p$, then $p \in M$,

without changing the concept of ideal. The proof is elementary.

Here is a general and useful remark about homomorphisms and their kernels: a necessary and sufficient condition that a homomorphism be a monomorphism (one-to-one) is that its kernel be $\{0\}$. Proof of necessity: if f is one-to-one and $f(p) = 0$, then $f(p) = f(0)$, and, therefore, $p = 0$. Proof of sufficiency: if the kernel of f is $\{0\}$ and if $f(p) = f(q)$, then $f(p + q) = f(p) + f(q) = 0$, so that $p + q = 0$, and this means that $p = q$.

Every example of a homomorphism (such as the ones we saw in §9) gives rise to an example of an ideal, namely its kernel. Thus if $f(p) = p \wedge p_0$ for every p, then the corresponding ideal consists of all those elements p for which $p \wedge p_0 = 0$, or, equivalently, $p \leq p_0'$. If f is defined on a field of subsets of X so that $f(P)$ is the value of the characteristic function of P at some particular point x_0 of X, then the corresponding ideal consists of all those sets P in the field that do not contain x_0. If, finally, the homomorphism f is induced by a mapping ϕ from a set X into a set Y, then the corresponding ideal consists of all those sets P in the domain of f that are disjoint from the range of ϕ.

There are examples of ideals for which it is not obvious that they are associated with some homomorphism. One such example is the class of all finite sets in the field of all subsets of a set. More generally, the class of all those finite sets that happen to belong to some particular field is an ideal in that field; a similar generalization is available for each of the following three examples. The class of all countable sets is an ideal in the field of all subsets of an arbitrary set; the class of all sets of measure zero is an ideal in the field of all measurable subsets of a measure space; and the class of all nowhere dense sets is an ideal in the field of all subsets of a topological space.

Every Boolean algebra B has a *trivial* ideal, namely the set $\{0\}$ consisting of 0 alone; all other ideals of B will be called *non-trivial*. Every Boolean algebra B has an *improper* ideal, namely B itself; all other ideals will be called *proper*. Observe that an ideal is proper if and only if it does not contain 1.

The intersection of every collection of ideals in a Boolean algebra B is again an ideal of B. It follows that if E is an arbitrary subset of B, then the intersection of all those ideals that happen to include E is an ideal. (There is always at least one ideal that includes E, namely the improper ideal B.) That intersection, say M, is the smallest ideal in B that includes E; in other words, M is included in every ideal that includes E. The ideal M is called the ideal *generated* by E. Thus, for example, if E is empty, then the ideal generated by E is the smallest possible ideal of B, namely the trivial ideal $\{0\}$. An ideal generated by a singleton $\{p\}$ is called a *principal* ideal; it consists of all the subelements of p.

The concepts of subalgebra and homomorphism are in a certain obvious sense self-dual; the concept of ideal is not. The dual concept is defined as follows. A Boolean *filter* in a Boolean algebra B is a subset N of B such that

(4) $$1 \in N,$$

(5) $$\text{if } p \in N \text{ and } q \in N, \text{ then } p \wedge q \in N,$$

(6) $$\text{if } p \in N \text{ and } q \in B, \text{ then } p \vee q \in N.$$

The condition (4) can be replaced by the condition that N be not empty. The condition (6) can be replaced by

$$\text{if } p \in N \text{ and } p \leqq q, \text{ then } q \in N.$$

Neither of these replacements will alter the concept being defined. The filter *generated* by a subset of B, and, in particular, a *principal* filter are defined by an obvious dualization of the corresponding definitions for ideals.

The relation between filters and ideals is a very close one. The fact is that filters and ideals come in dual pairs. This means that there is a one-to-one correspondence that pairs each ideal to a filter, its dual, and by means of which every statement about ideals is immediately translatable to a statement about filters. The pairing is easy to describe. If M is an ideal, write $N = \{p : p' \in M\}$, and, in reverse, if N is a filter, write $M = \{p : p' \in N\}$. It is trivial to verify that this construction does indeed convert an ideal into a filter, and vice versa.

Ideals and filters have their complete versions. A *complete* ideal is an ideal M such that the supremum of all its subsets exists and belongs to M. The importance of complete ideals is that the kernel of every complete homomorphism on a complete algebra is a complete ideal. Complete ideals, nevertheless, do not play a large role; the reason is that every complete ideal is principal. (Proof: if M is a complete ideal, consider $\bigvee M$.)

Exercises

(1) Is every finitely generated ideal a principal ideal? (An ideal is *finitely generated* if it is generated by a finite set.)

(2) Prove that if M is an ideal in a Boolean algebra and N is its associated (dual) filter, then the set-theoretic union $M \cup N$ is a subalgebra.

§ 12. The homomorphism theorem

An ideal is *maximal* if it is a proper ideal that is not properly included in any other proper ideal. Equivalently, to say that M is a maximal ideal in B means that $M \neq B$, and, moreover, if N is an ideal such that $M \subset N$, then either $N = M$ or $N = B$. Examples: the trivial ideal is maximal in 2; the ideals, in fields of sets, defined by the exclusion of one point are maximal.

Maximal ideals are characterized by a curious algebraic property.

LEMMA 1.   *An ideal M in a Boolean algebra B is maximal if and only if either $p \in M$ or $p' \in M$, but not both, for each p in B.*

Proof.   Assume first that, for some p_0 in B, neither $p_0 \in M$ nor $p_0' \in M$; it is to be proved that M is not maximal. Let N be the set of all elements of the form $p \vee q$, where $p \leqq p_0$ and $q \in M$. Direct verification shows that N is an ideal including M and containing p_0; in fact, N is exactly the ideal generated by $M \cup \{p_0\}$. It follows that $N \neq M$; it remains only to prove that $N \neq B$. This follows from the fact that $p_0' \in' N$. Indeed, if $p_0' = p \vee q$, with $p \leqq p_0$ and $q \in M$, then (form the meet of both sides with p_0') $p_0' = q$, contradicting the assumption $p_0' \in' M$.

The converse is easier. Assume that always either $p \in M$ or $p' \in M$, and suppose that N is an ideal properly including M; it is to be proved that $N = B$. Since $N \neq M$, there is an element p in N that does not belong to M. The assumption implies that $p' \in M$, and therefore $p' \in N$; since N is an ideal, it follows that $p \vee p' \in N$.

The definition of ideals was formulated so as to guarantee that the kernel of every homomorphism is an ideal; since a homomorphism never maps 1 onto 0, the kernel of every homomorphism is even a proper ideal. It is natural and important to raise the converse question: is every proper ideal the kernel of some homomorphism? For maximal ideals the answer is easily seen to be yes. Suppose, indeed, that M is a maximal ideal in B, and write $f(p) = 0$ or 1 according as the element p of B belongs to M or not. In view of Lemma 1, the definition of f can also be formulated this way: $f(p) = 0$ or 1 according as $p \in M$ or $p' \in M$. A straightforward verification, based on Lemma 1, shows that f is a homomorphism from B to 2; the kernel of f is obviously M. What we have proved in this way is a very special case of the following result, known as the *homomorphism theorem*.

THEOREM 2. *Every proper ideal is the kernel of some epimorphism.*

Proof. The simplest way to settle the matter is to refer to the theory of rings. If B is a Boolean ring and M is a proper ideal in B, then the quotient B/M is a ring. The idempotence of the elements of B implies the same for B/M. The desired epimorphism is the so-called *natural* or *canonical* mapping, or *projection*, from B onto B/M; it associates with each element of B the equivalence class (or coset) of that element modulo M. The Boolean algebra B/M is called the *quotient* of B modulo M.

Associated with the homomorphism theorem there is a cluster of results of the universal algebraic kind, some of which we now proceed to state.

Suppose that M is a proper ideal in a Boolean algebra B, write $A = B/M$, and let f be the projection from B to A. The mapping that associates with every ideal N in A the set $f^{-1}(N)$ in B is a one-to-one correspondence between all the ideals in A and all the ideals that include M in B. The images of the trivial ideal and of A under this correspondence are M and B, respectively. If $N_1 \subset N_2$, then $f^{-1}(N_1) \subset f^{-1}(N_2)$. If f_0 is a homomorphism from B to a Boolean algebra A_0, say, and if the kernel M_0 of f_0 includes M, then there exists a unique homomorphism g from A to A_0 such that $f_0 = g \circ f$.

The proofs of all these assertions are the same for Boolean algebras as for other algebraic structures (such as groups and rings); the words may change but the ideas stay the same. It is not worth while to record the proofs here; the interested reader should have no serious difficulty in reconstructing them.

A Boolean algebra is called *simple* if it has no non-trivial proper ideals. Simplicity is a universal algebraic concept, but, as it turns out, in the context of Boolean algebras it is not a fruitful one. The reason is that there is just exactly one simple algebra, namely 2. Clearly 2 is simple. If, conversely, B is simple, and if p is a non-zero element of B, then the principal ideal generated by p must be improper, which can happen only if $p = 1$. In other words, B is such that if an element of B is not 0, then it is 1; this means that $B = 2$.

The correspondence between the ideals of a quotient algebra and the ideals of its "numerator" shows that the quotient algebra is simple if and only if its "denominator" (the ideal) is maximal. For Boolean algebras this means, via the preceding paragraph, that a quotient algebra is equal to 2 if and only if its denominator is maximal. This is a useful observation, but not a profound one; it is not even as deep as the general homomorphism theorem.

Exercises

(1) Prove that every Boolean ring without a unit can be embedded as a maximal ideal into a Boolean ring with a unit. To what extent is the extension unique? (Cf. Exercise 1.1.)

(2) Prove that two epimorphisms with the same domain and the same kernel have isomorphic ranges.

(3) Prove that the quotient of a complete algebra by a complete ideal is complete.

§ 13. Boolean σ-algebras

Between Boolean algebras and complete Boolean algebras there is room for many intermediate concepts. The most important one is that of a Boolean σ-algebra; this means, by definition, a Boolean algebra in which every countable set has a supremum (and therefore, of course, an infimum). Similarly a field of sets is a σ-field if it is closed under the formation of countable unions (and intersections).

It is a routine matter to imitate the entire algebraic theory developed so far for the two extremes (Boolean algebras and complete algebras) in the intermediate case of σ-algebras. Thus a σ-subalgebra of a σ-algebra is one that

is closed under the formation of countable suprema; a
σ-subalgebra of a σ-field of sets is called a *σ-subfield*. The
definition of the σ-subalgebra generated by a set is an
equally obvious modification of the concepts treated before.

Continuing in the same spirit, we define a *σ-homomorphism*
as a homomorphism that preserves all the countable suprema
(that is, the suprema of all the countable sets) that happen
to exist. A *free σ-algebra* is defined the same way as a free
Boolean algebra except that all the algebras and homomorphisms that enter the definition are now required to be
σ-algebras and σ-homomorphisms. (The problem of the existence of σ-algebras free on sets of generators of arbitrary
cardinality will be attacked later.)

A *σ-ideal* is an ideal closed under the formation of
countable suprema. The kernel of a σ-homomorphism on a
sigma algebra is a σ-ideal, and, conversely, every proper
σ-ideal is the kernel of a σ-epimorphism. The latter assertion is the only thing that requires proof; here is how it goes.

Suppose that B is a σ-algebra and M is a σ-ideal in B.
Write $A = B/M$ and let f be the projection of B onto A. We
shall prove that A is a σ-algebra and f is a σ-homomorphism.
The two assertions can be treated simultaneously by proving
that if $\{q_n\}$ is a sequence of elements in B, then the sequence $\{f(q_n)\}$ has a supremum in A, and, in fact,

$$\bigvee_n f(q_n) = f\left(\bigvee_n q_n\right).$$

Write $f(q_n) = p_n$, $\bigvee_n q_n = q$, and $f(q) = p$. Clearly $p_n \leqq p$
for all n; it is to be proved that if $p_n \leqq s$ for all n, then
$p \leqq s$. Let t be an element of B such that $f(t) = s$. Since
$f(q_n) \leqq f(t)$ for all n, or $f(q_n - t) = 0$, it follows that
$\bigvee_n (q_n - t) \in M$ (recall that M is a σ-ideal). This implies
that $f(q) \leqq f(t)$, as promised.

The simplest way to be a σ-algebra is to be complete. There are other ways. The countable-cocountable algebra of every set is a σ-algebra that is not complete, unless the underlying set is countable. (Observe, by the way, that the class of all countable sets in this algebra is a non-trivial maximal ideal.) The most famous and useful incomplete σ-algebras arise in topological spaces. A *Borel set* in a topological space is, by definition, a set belonging to the σ-field generated by the class of all open sets (or, equivalently, by the class of all closed sets). There is also an interesting σ-ideal that can be defined in topological terms. A subset of a topological space is *meager* (Bourbaki) if it is the union of countably many nowhere dense sets. (In classically clumsy nomenclature meager sets are called sets of the *first category*.) The class of all meager subsets of a topological space X is a σ-ideal in $\mathscr{P}(X)$, and, consequently, the class of all meager Borel sets is a σ-ideal in the σ-algebra of all Borel sets.

The following celebrated result, known as the *Baire category theorem*, is needed on most occasions when meager sets occur. The corresponding result for complete metric spaces, instead of compact Hausdorff spaces, is of importance in analysis.

THEOREM 3. *A meager open set in a compact Hausdorff space is empty.*

Proof. Suppose that U is a non-empty open set and that $\{S_n\}$ is a sequence of nowhere dense sets; we shall show that U contains at least one point that does not belong to any S_n. Let V_1 be a non-empty open set such that $V_1^- \subset U$, and let U_1 be a non-empty open subset of V_1 such that $U_1 \cap S_1 = \emptyset$. (This uses, of course, the assumption that S_1 is nowhere dense.) Let V_2 be a non-empty open set such that $V_2^- \subset U_1$, and let U_2 be a non-empty open subset of V_2 such that $U_2 \cap S_2 = \varnothing$. The inductive procedure so begun yields a decreasing sequence $\{U_k\}$ of open sets such that $\cap_k U_k = \cap_k U_k^- \neq \varnothing$, and such that $\cap_k U_k$ is disjoint from each S_n.

The ideal of meager sets makes contact with an earlier construction in a somewhat surprising way.

THEOREM 4. *Suppose that B is the σ-field of Borel sets and M is the σ-ideal of meager Borel sets in a compact Hausdorff space X. Corresponding to each set S in B there exists a unique regular open set $f(S)$ such that $S + f(S) \in M$. The mapping f is a σ-homomorphism from B onto the algebra A of all regular open sets with kernel M, so that A is isomorphic to B/M.*

Proof. Using the ordinary language of ideal theory we shall say that S is *congruent* to U modulo M, and we shall write $S \equiv U$ (mod M), in case $S + U \in M$. A subset S of X is said to have the *Baire property* if it is congruent to some open set. (By "congruent" in the course of this proof we shall always understand "congruent modulo M".) Clearly every open set has the Baire property, and the class of sets with the Baire property is closed under the formation of countable unions. If $S \equiv U$, where U is open, then $S \equiv U^-$ (see Lemma 4.5), and therefore $S' \equiv U^\perp$. This implies that the class of sets with the Baire property is a σ-field that includes all open sets. Conclusion: every Borel set has the Baire property.

If U is open, then $U \subset U^{\perp\perp}$ (Lemma 4.2)). Since $U^{\perp\perp} \subset U^-$, it follows that $U \equiv U^{\perp\perp}$. From this, together with the result of the preceding paragraph, it follows that every Borel set is congruent to some regular open set.

If U and V are congruent regular open sets, then, since $U \equiv U^-$ and $V \equiv V^-$, it follows that both $U + V^-$ and $U^- + V$ are congruent to $\emptyset$. It follows that both $U - V^-$ and $V - U^-$ are meager; since they are also open, the Baire category theorem implies that $U \subset V^-$ and $V \subset U^-$. This implies that the closures of U and V are the same, and hence, by their regular open character, so also are U and V.

We have seen above that if $S \equiv U$, where U is open, then $S' \equiv U^{\perp}$, and if $S_n \equiv U_n$, $n = 1, 2, \cdots$, where again the U_n's are open, then $\bigcup_n S_n \equiv (\bigcup_n U_n)^{\perp\perp}$. These two assertions mean just that f is a σ-homomorphism. The assertion about the kernel of f is true by definition; the onto character of f follows from the fact that if S is a regular open set, then $f(S) = S$. The proof of the theorem is complete.

One surprising aspect of the theorem is that the quotient of a σ-algebra by a σ-ideal, which is necessarily a σ-algebra itself, turns out to be a complete algebra. This is a special dividend; it is not to be expected in every case.

It is tempting, but not particularly profitable, to define classes of Boolean algebras depending on other cardinal numbers the same way as σ-algebras depend on $\aleph_0$. The situation is analogous to the various generalizations of compactness depending on cardinal numbers. The questions undeniably exist, the answers are sometimes easy and sometimes not, and the answers are sometimes the same as for the ungeneralized concepts and sometimes not. In all cases, however, and in Boolean algebra as well as in topology, the generalized theory has much more the flavor of cardinal number theory than of the subject proper. The interested reader should have no trouble in reconstructing the basic theory for himself. The problem is, given an infinite cardinal m, to define and to study m-algebras, m-fields, m-subalgebras, m-subfields, m-homomorphisms, free m-algebras, m-ideals, m-filters, etc. For complicated historical reasons the symbol "$\aleph_0$" is always replaced by "σ" in such contexts, so that, for instance, $\aleph_0$-algebras are the same as the σ-algebras that constituted the main subject of this section.

Exercises

(1) Give an example of an incomplete σ-field.

(2) Define *σ-regular subalgebras*, in analogy with the regular subalgebras introduced in Exercise 8.6, and investigate whether the results of Exercises 8.6 and 8.7 extend to this concept.

(3) Is every set with the Baire property a Borel set?

(4) Can the ideal of meager sets be maximal?

(5) Prove the Baire category theorem for locally compact Hausdorff spaces.

(6) Is the homomorphism f described in Theorem 3 complete?

(7) Prove that if A is a Boolean σ-algebra, and if p is an element of the σ-algebra generated by a subset E of A, then E has a countable subset D such that p belongs to the σ-subalgebra generated by D.

§ 14. The countable chain condition

The algebraic behavior of the regular open algebra of a topological space X reflects, at least in part, the topological properties of X. One particular topological property of X, namely the possession of a countable base, has important algebraic repercussions, which we now proceed to study.

A Boolean algebra A is said to satisfy the *countable chain condition* if every disjoint set of non-zero elements of A is countable. (Two elements p and q of a Boolean algebra are *disjoint* if $p \wedge q = 0$; a set E is disjoint if every two distinct elements of E are disjoint.) The regular open algebra of a space with a countable base does satisfy the countable chain condition. Proof: select a countable base, and, given a disjoint class of non-empty regular open sets, find in each one a set of the base. An algebra satisfying the countable chain condition is sometimes called *countably decomposable*.

LEMMA 1. *A Boolean algebra A satisfies the countable chain condition if and only if every set E in A has a countable subset D such that D and E have the same set of upper bounds.*

Proof. Assume first that the condition is satisfied and suppose that E is a disjoint set of non-zero elements of A. Let D be a countable subset of E with the same set of upper bounds. If E had an element not in D, the complement of such an element would be an upper bound of D but not of E. Conclusion: $E = D$, and therefore E is countable.

To prove the converse, assume now that the countable chain condition is satisfied and let E be an arbitrary subset of A. Let M be the ideal generated by E; the elements of M are just those elements of A that can be dominated by the supremum of some finite subset of E. It follows that M and E have the same set of upper bounds. Apply Zorn's lemma to find a maximal disjoint set, say F, of non-zero elements of M. Reasoning as in the preceding paragraph, we infer that F and M have the same set of upper bounds. Since the countable chain condition holds, the set F is countable. Since each of the countably many elements of F is dominated by the supremum of some finite subset of E, the union, say D, of all these finite sets is a countable subset of E with the same set of upper bounds.

COROLLARY. *A Boolean σ-algebra that satisfies the countable chain condition is complete.*

Proof. Every countable supremum is formable by definition; by Lemma 1 every conceivable supremum coincides with some countable one.

The countable chain condition got its name from its close relation to a condition in which ascending chains do explicitly occur. An *ascending well-ordered chain* in a Boolean algebra A is a function that associates with each element α of some well-ordered set an element p_α of A so that $p_\alpha \leqq p_\beta$ whenever $\alpha \leqq \beta$. The chain is *strictly ascending* if $p_\alpha \neq p_\beta$ whenever $\alpha < \beta$, and the chain is called countable in case the set of indices is countable.

LEMMA 2. *If a Boolean algebra A satisfies the countable chain condition, then every strictly ascending well-ordered chain in A is countable.*

Proof. Suppose that $\{p_a\}$ is a strictly ascending well-ordered chain, and assume, with no loss of generality, that the index set consists of all ordinal numbers less than some particular infinite ordinal number, say γ. Write $q_a = p_{a+1} - p_a$ whenever $a + 1 < \gamma$, and let E be the set of q_a's. The cardinal number of E is the same as that of γ. The elements of E are distinct from 0, since $p_{a+1} \neq p_a$. If $a < \beta$ and $\beta + 1 < \gamma$, then $p_{a+1} \leqq p_\beta$, and therefore

$$q_a \wedge q_\beta = p_{a+1} \wedge p_a' \wedge p_{\beta+1} \wedge p_\beta' \leqq p_{a+1} \wedge p_\beta' = 0.$$

In other words E is a disjoint set of non-zero elements and therefore countable; it follows that the given chain is countable.

In a Boolean σ-algebra the converse of Lemma 2 is also true.

LEMMA 3. *If every strictly ascending well-ordered chain in a Boolean σ-algebra A is countable, then A satisfies the countable chain condition.*

Proof. If the conclusion is false, then there exists a disjoint set E of cardinal number $\aleph_1$ consisting of non-zero elements of A. Establish a one-to-one correspondence between E and the set of all ordinal numbers less than Ω (the first uncountable ordinal number). Let p_a be the element of E corresponding to a $(a < \Omega)$. Since the number of predecessors of a is countable, it makes sense to write $q_a = \bigvee_{\beta < a} p_a$ for each a. Since $\{q_a\}$ is a strictly ascending well-ordered chain (strictness follows from the disjointness of E) the hypothesis of the lemma leads to the contradictory conclusion that Ω is countable.

Exercises

(1) If the regular open algebra of a topological space satisfies the countable chain condition, does it follow that the space has a countable base?

(2) Show that the converse of Lemma 2 is false. (Hint: consider the finite-cofinite algebra of an uncountable set.)

(3) Show that the countable chain condition is not preserved by homomorphisms. (Hint: consider the algebra of all subsets of a countable set modulo the ideal of all finite sets. For ease in manipulation, let the countable set be the set of all rational numbers, and, for each real number t, find a set of rational numbers that has t as its unique limit point.)

(4) Prove that a Boolean algebra A satisfies the countable chain condition if and only if every subset E of A that has a supremum has a countable subset D such that D has a supremum and $\bigvee D = \bigvee E$. (Hint: every disjoint set of non-zero elements can be embedded in a maximal set of that kind and that maximal set necessarily has a supremum, namely 1.)

§ 15. Measure algebras

A *measure* on a Boolean algebra A is a non-negative real-valued function μ on A such that whenever $\{p_n\}$ is a disjoint sequence of elements of A with a supremum p in A, then $\mu(p) = \sum_n \mu(p_n)$. The principal condition that this definition imposes is called countable additivity, so that a measure can be described as a non-negative and countably additive function on a Boolean algebra.

The concept just defined is the most useful one of a large collection of related concepts. Sometimes the word "measure" is applied to countably additive functions whose values are arbitrary real numbers, or complex numbers, or elements of much more general algebraic structures. Sometimes the condition of countable additivity is relaxed to *finite additivity* .(The meaning of this phrase should be obvious. Note that μ is finitely additive if and only if $\mu(p \vee q) = \mu(p) + \mu(q)$ whenever p and q are disjoint.) If ever we need to make use of such generalized concepts we shall refer to them by appropriately qualifying "measure". (Thus, for instance, we may speak of a complex-valued finitely additive measure.)

Examples of measures are easy to obtain. For a combinatorial example consider the field $\mathcal{P}$ (X) of all subsets of a finite set X and, for each P in $\mathcal{P}$ (X), define $\mu(P)$ to be the number of points in P. Many examples occur in analysis; perhaps the simplest is Lebesgue measure on the algebra of Lebesgue measurable subsets of the closed unit interval. A more sophisticated example is given by Haar measure on, say, the algebra of Borel sets in a compact topological group.

A measure μ is *normalized* if $\mu(1) = 1$; it is *positive if* 0 is the only element at which μ takes the value 0.

LEMMA 1. *Let ν be a normalized measure on a Boolean σ-algebra B and let M be the set of all those elements q of B for which ν $(q) = 0$. The set M is a proper σ-ideal in B. If $A = B/M$ and if f is the projection of B onto A, then there exists a unique measure μ on A such that $\mu(f(q)) = \nu$ (q) for all q on B; the measure μ is normalized and positive.*

Proof. We shall prove the existence of μ and its positiveness; the remaining assertions of the lemma are

trivial. Given p in A, find q in B with $f(q) = p$ and write
$\mu(p) = \nu(q)$. If $f(q_1) = f(q_2)$, then $f(q_1 + q_2) = 0$, so that
$q_1 + q_2 \in M$ or $\nu(q_1 + q_2) = 0$. This implies that $\nu(q_1) = \nu(q_2)$,
and hence that the definition of μ is unambiguous. To
prove that μ is countably additive, suppose that $\{p_n\}$ is a
disjoint sequence in A and let $\{q_n\}$ be a sequence in B such
that $f(q_n) = p_n$. The sequence $\{q_n\}$ may not be disjoint, but
it can be disjointed. More precisely, there exists a disjoint
sequence $\{r_n\}$ with $f(r_n) = p_n$, obtained as follows:

$$r_1 = q_1 ,$$

$$r_2 = q_2 - q_1 ,$$

$$r_3 = q_3 - (q_1 \vee q_2) ,$$

$$r_4 = q_4 - (q_1 \vee q_2 \vee q_3) ,$$
$$\cdot\ \cdot\ \cdot\ \cdot\ \cdot$$

A routine examination proves that $f(r_n) = p_n$; once that is
known, the countable additivity of μ becomes an obvious
consequence of the corresponding property of ν. To prove
that μ is positive, suppose that $\mu(p) = 0$ for some p in A.
It follows that $\nu(q) = 0$ whenever $f(q) = p$, and hence that
$q \in M$ whenever $f(q) = p$. This implies that $p = 0$ whenever
$\mu(p) = 0$.

Lemma 1 says that under certain conditions measures
can be transferred to quotient algebras. The reverse
always works; a measure on a quotient can always be
lifted to its numerator.

LEMMA 2. *Let f be a Boolean σ-epimorphism from a
σ-algebra B to a σ-algebra A, and let μ be a normalized
measure on A. If $\nu(q) = \mu(f(q))$ for every q in B, then ν is
a normalized measure on B. The kernel of f is included in
the set of all those elements q of B for which $\nu(q) = 0$;*

the kernel coincides with that set if and only if the measure
μ is positive.

The proofs of all the assertions of the lemma are immediate from the definitions.

It is sometimes useful to consider a measure as an
intrinsic part of the Boolean algebra it is defined on. The
appropriate definition is that of a *measure algebra*, defined
as a Boolean σ-algebra *A* together with a positive, normalized measure μ on *A*. If *A* is not required to be a σ-algebra,
but just a Boolean algebra, and if, correspondingly, μ is
required to be only finitely additive, we may speak of a
finitely additive measure algebra.

The theory of measure algebras has several points of
contact, in both form and content, with the topological and
algebraic results of the preceding two sections. Countability,
for instance, enters through the essential countability properties of real numbers, as follows.

LEMMA 3. *Every finitely additive measure algebra*
satisfies the countable chain condition.

Proof. A disjoint set of non-zero elements cannot contain, for any positive integer n, as many as n elements of
measure greater than $1/n$.

COROLLARY. *Every measure algebra is complete.*

Proof. Apply the preceding lemma and the corollary of
Lemma 14.1.

The reduced Borel algebra (Borel sets modulo meager
Borel sets) and the reduced measure algebra (Borel sets

modulo Borel sets of measure zero) of the unit interval
have much in common. Both algebras are obtained by reducing
an incomplete σ-field modulo a σ-ideal; both algebras satisfy
the countable chain condition and therefore (Corollary of
Lemma 14.1) both algebras are complete; and, incidentally,
both algebras are non-atomic. (The proof of the latter
assertion is a trivial consequence of Theorem 4 (p. 58) for
the reduced Borel algebra; for the reduced measure algebra
it requires an elementary measure-theoretic argument.) No
property of Boolean algebras that we have encountered so
far is sharp enough to tell these two algebras apart; for all
we know they are isomorphic. We conclude this section by
showing that they are not. (Note incidentally that Borel
sets modulo Borel sets of measure zero and Lebesgue
measurable sets modulo Lebesgue measurable sets of
measure zero are the same. This depends on the fact that
every Lebesgue measurable set differs from some Borel
set in a set of measure zero only.)

LEMMA 4. *Every measure on the reduced Borel algebra
of the closed unit interval is identically zero.*

Proof. Let B be the σ-field of Borel sets in $[0, 1]$, and
let M be the σ-ideal of meager sets in B. Write $A = B/M$,
and let f be the projection of B onto A. If there were a
non-zero measure μ on A, we could assume, with no loss of
generality, that μ is normalized. An application of Lemma 2
yields a normalized measure ν on B that vanishes on
every meager Borel set. By a standard construction (cover
the rational points with open intervals of small measure),
the interval is the union of two disjoint Borel sets S and T
such that S is meager and $\nu(T) = 0$. Since $f(S) = 0$, so that
$f([0, 1]) = f(T)$, it follows that $\nu([0, 1]) = 0$, a contradiction.

Exercises

(1) State and prove the analogues of Lemmas 1 and 2 for finitely additive measures.

(2) If A is a measure algebra with measure μ, and if $d(p, q) = \mu(p + q)$, then d is a metric on A; prove that with respect to this metric A is a complete metric space.

§ 16. Atoms

The most natural field of subsets of a set is the field of all its subsets. Does that field have a simple algebraic characterization? The answer is yes; the purpose of this section is to exhibit such a characterization.

An *atom* of a Boolean algebra is an element that has no non-trivial proper subelements. Better: q is an atom if $q \neq 0$ and if there are only two elements p such that $p \leqq q$, namely 0 and q. A typical example of an atom is a singleton in a field of sets. A Boolean algebra is *atomic* if every non-zero element dominates at least one atom. A Boolean algebra is *non-atomic* if it has no atoms. (Note that these two concepts are not just the negations of one another.) A field of sets is usually (but not always) atomic: the field of all subsets, or the finite-cofinite algebra of a set are obvious examples. A counterexample is the field generated by half-closed intervals in the line; it is non-atomic. The regular open algebra of a topological space X is quite likely to be non-atomic; the absence of separation axioms and the presence of isolated points, however, is likely to introduce atoms.

LEMMA 1. *In an atomic algebra every element is the supremum of the atoms it dominates.*

Proof. The statement of the theorem is intended to convey the information that the supremum in question always exists (without any assumption of completeness). Observe also that even the zero element does not have to be excluded from the statement. Now for the proof itself: begin with the trivial comment that each element p is an upper bound of the set, say E, of the atoms that it dominates. It is to be proved that if r is an arbitrary upper bound of E, then $p \leqq r$. Assume that, on the contrary, $p - r \neq 0$. It follows (from the assumption of atomicity) that there exists an atom q with $q \leqq p - r$. Since $p - r \leqq p$, the atom q belongs to E; since, however, $q \wedge r \leqq (p - r) \wedge r$, this contradicts the fact that r is an upper bound of E.

THEOREM 5. *A necessary and sufficient condition that a Boolean algebra A be isomorphic to the field of all subset of some set is that A be complete and atomic, or, alternatively, that A be complete and completely distributive.*

Proof. The necessity of either pair of conditions is obvious. Suppose therefore that A is complete and atomic, and let X be the set of all atoms of A. For each p in A let $f(p)$ be the set of all those elements q of X for which $q \leqq p$. Trivially $f(p_1) \cup f(p_2) \subset f(p_1 \vee p_2)$. To prove the reverse inclusion, suppose that $q \in X$ and $q \leqq p_1 \vee p_2$. It follows that $q = q \wedge (p_1 \vee p_2) = (q \wedge p_1) \vee (q \wedge p_2)$. At least one of $q \wedge p_1$ and $q \wedge p_2$ must be different from 0, and that one, since q is an atom, must be equal to q; this proves that $q \in f(p_1) \cup f(p_2)$. We know therefore that f preserves joins, and therefore, in particular, $f(p) \cup f(p') = f(1) = X$ for every p. Since $f(p)$ and $f(p')$ are obviously disjoint, it follows that f preserves complementation also.

In other words f is a homomorphism from A to a field of subsets of X. If E is an arbitrary subset of X, then, by completeness, E has a supremum in A. If $p = \bigvee E$, then $f(p) = E$ so that f is an epimorphism from A to $\mathscr{P}(X)$. All that is needed to complete the proof is to show that f is one-to-one, that is, that the kernel of f is trivial. This follows from Lemma 1: since $\bigvee f(p) = p$, the only way $f(p)$ can be 0 is to have $p = 0$.

The sufficiency of the second pair of conditions is proved by showing that if A is complete and completely distributive, then A is atomic. To apply complete distributivity, write $I = A$, $J = \{+1, -1\}$, and $p(i, j) = i$ or i' according as $j = +1$ or $j = -1$, for each i in I. Since $\bigvee_{j \in J} p(i, j) = 1$ for every i, it follows from (7.1) that

$$\bigvee_{a \in J^I} \bigwedge_{i \in I} p(i, a(i)) = 1,$$

and consequently, by Lemma 7.6,

$$\bigvee_{a \in J^I} (r \wedge \bigwedge_{i \in I} p(i, a(i))) = r$$

for every r in A. The proof will be completed by showing that every non-zero element of the form $\bigwedge_{i \in I} p(i, a(i))$ is an atom of A. Suppose accordingly that $q = \bigwedge_{i \in I} p(i, a(i)) \neq 0$ and that r is a non-zero element of A such that $r \leq q$. Since $q \leq p(r, a(r))$, two things follow: (1) $a(r) = +1$, for otherwise $r \leq r'$, contradicting the fact that $r \neq 0$, and, therefore, (2) $q \leq r$. This implies that $r = q$, so that q is indeed an atom, and the proof is complete.

Exercises

(1) Prove without using Theorem 5 that every finite Boolean algebra is atomic. (Since a finite algebra is obviously complete and completely distributive, Theorem 5

could be used. The conclusion is too elementary to deserve such a relatively high-powered treatment.)

(2) Prove that the total number of elements in every finite Boolean algebra is a power of 2, and that two finite Boolean algebras with the same number of elements must be isomorphic.

(3) Prove that a finitely generated Boolean algebra is finite, and, in fact, the number of elements in an algebra with n generators is $\leq 2^{2^n}$. (Hint: if the generators are r_i, $i = 1, \cdots, n$, write $J = \{+1, -1\}$, and put $p(i, j) = r_i$ or r_i' according as $j = +1$ or $j = -1$. The non-zero elements of the form $\bigwedge_i p(i, a(i))$, where $a \in J^I$, are atoms.)

(4) If p is a non-zero element of an atomic Boolean algebra A, then there exists a 2-valued homomorphism f on A such that $f(p) = 1$.

(5) Characterize the topological spaces whose regular open algebra is (1) atomic, (2) non-atomic.

(6) Prove that the mapping f defined in the proof of Theorem 5 is a complete homomorphism.

(7) Does the set of all atoms in a Boolean algebra always have a supremum?

§ 17. Boolean spaces

We know by now that not every Boolean algebra is isomorphic to the field of all subsets of some set. In the next section we shall prove that every Boolean algebra is isomorphic to some field of subsets of some set. In order to get a usable description of what kind of fields and what kind of sets are needed, we proceed now to introduce a rather special category of topological spaces.

A *Boolean space* is a totally disconnected compact Hausdorff space. There are several possible definitions of total disconnectedness, but, as it turns out, they are all equivalent for compact Hausdorff spaces. The most convenient definition for our algebraic purposes is the one that demands that the clopen sets constitute a base. Explicitly: a Boolean space is a compact Hausdorff space with the property that every open set is the union of those simultaneously closed and open sets that it happens to include.

For Boolean spaces, as for every topological space, it is true that the class of all clopen sets is a field. The field of all clopen sets in a Boolean space X is called the *dual algebra* of X.

The simplest Boolean spaces are the finite discrete spaces. Since every subset of such a space is clopen, the dual algebra of each finite Boolean space is a finite Boolean algebra. Since every finite Boolean algebra is isomorphic to the field of all subsets of some (necessarily finite) set (see §16), it follows that every finite Boolean algebra is isomorphic to the dual algebra of some discrete Boolean space.

A less trivial collection of examples consists of the one-point compactifications of infinite discrete spaces. Explicitly, suppose that a set X with a distinguished point x_0 is topologized as follows: every subset of the complement of $\{x_0\}$ is open, but a set containing x_0 is open if and only if its complement is finite. It is easy to verify that the space X so defined is Boolean; a subset of X is clopen if and only if it is either a finite subset of $X - \{x_0\}$ or a cofinite subset (of X) containing x_0. The dual algebra of X is isomorphic to the finite-cofinite algebra of $X - \{x_0\}$.

The set 2 is a Boolean algebra; from now on it will be
convenient to construe it as a topological space as well,
endowed with the discrete topology. For an arbitrary set I,
the set 2^I of all functions from I into 2 (equivalently: the
Cartesian product of copies of 2, one for each element of
I) is a topological space (product topology); it is well
known that that space is compact and Hausdorff (Tychonoff's
theorem). We shall denote the value of a function x in 2^I
at an element i of I by x_i. The sets of the form
$\{x \in 2^I : x_i = \delta\}$, where $i \in I$ and $\delta \in 2$, constitute a subbase
for 2^I; finite intersections of them constitute a base. Since
the complement of each set of the indicated form is another
set of the same form, so that each such set is clopen, it
follows that 2^I is a Boolean space. In the sequel these
particular Boolean spaces will be called *Cantor spaces*.

The following somewhat technical result is useful in
the study of Boolean spaces.

LEMMA 1. *If X is a compact Hausdorff space and if A
is a separating field of clopen subsets of X, then X is a
Boolean space and A is the field of all clopen subsets of
X. (To say that A is separating means that for every pair
of distinct points x and y in X there exists a set P in A
with $x \in P$ and $y \in P'$.)*

Proof. The fact that A separates points implies that A
separates points and closed sets. This involves a standard
compactness argument. Suppose, indeed, that F is a closed
set and x is a point not in F. Separate each point of F
from x by a suitable set in A. Compactness yields a finite
cover of F by sets in A none of which contains x; their
union is a set in A that separates x from F. (The union is
in A because A is a field.)

The result of the preceding paragraph can be rephrased
by saying that A is a base for X; this already implies that
X is Boolean. It follows that every clopen set in X is a

finite union of sets of A (because it is both open and compact). Since A is closed under the formation of finite unions, the proof is complete.

COROLLARY. *If a field of clopen subsets of a compact Hausdorff space is a base, then the space is Boolean and the field contains all clopen sets.*

LEMMA 2. *Every closed subset Y of a Boolean space X is a Boolean space with respect to the topology it inherits from X. Every clopen set in Y is the intersection of Y with some clopen subset of X.*

Proof. The first statement is obvious: if the clopen sets form a base in X, their intersections with Y do the same for Y. If Q is clopen in Y, then it is open in Y, and, therefore, there exists an open set U in X such that $Q = Y \cap U$. The clopen subsets of U in X cover the closed set Q, and, therefore, by compactness, there exists a finite class of clopen subsets of U whose union, say P, covers Q. Since $Q \subset P \subset U$ and $Y \cap U = Q$, it follows that $Y \cap P = Q$.

Exercises

(1) Let X be the set of all ordinal numbers up to and including some particular one. The set X is ordered (by magnitude), and, as such, has a natural topology, namely the one for which the open intervals constitute a base. Prove that X is a Boolean space.

(2) Let X be the perimeter of a circle in the Cartesian plane. Order X as follows: (x_1, x_2) precedes (y_1, y_2) if and only if either $x_1 < y_1$, or (in case $x_1 = y_1$) $x_2 < y_2$. (This is known as the lexicographic ordering.) Endow X with the

order topology (as defined in a similar situation in Exercise 1 above). Prove that X is a Boolean space whose dual algebra is the field of half-closed intervals in the closed unit interval (see §3).

(3) Prove that if I is countably infinite, then the Cantor space 2^I is homeomorphic to the Cantor middle-third set.

(4) Prove that the Stone-Cech compactification of an infinite discrete space X is a Boolean space whose dual algebra is isomorphic to $\mathscr{P}(X)$.

(5) Prove that a compact.Hausdorff space is a Boolean space if and only if all its components are singletons.

(6) If I is an infinite set, of power m, say, what is the cardinal number of the dual algebra of the Cantor space 2^I?

(7) Prove that the dual algebra of every Cantor space satisfies the countable chain condition. (Hint: regard the Cantor space as a topological group and use Haar measure. Note that this gives a solution of Exercise 14.1.)

(8) Is the Cartesian product of a family of Boolean spaces a Boolean space with respect to the usual product topology?

(9) Imitating the definition of a free Boolean algebra, define the concept of a free Boolean space, and prove that every finite Boolean space is free but there are no infinite free Boolean spaces.

§ 18. The representation theorem

If a Boolean algebra A is a field of subsets of a set X, and, in particular, if it is the dual algebra of a Boolean space X, then the points of X serve to define 2-valued homomorphisms on A (see §9). This comment suggests that if we start with a Boolean algebra A and seek to represent it as the dual of some Boolean space X, a reasonable place to conduct the search for points suitable to make up X is among the 2-valued homomorphisms of A. The suggestion would be impractical if it turned out that A has no 2-valued homomorphisms. Our first result along these lines is that there is nothing to fear; there is always a plethora of 2-valued homomorphisms.

LEMMA 1. *For every non-zero element p of every Boolean algebra A there is a 2-valued homomorphism x on A such that x(p) = 1.*

Proof. In view of the results of §12, the conclusion can be rephrased as follows: there exists a maximal ideal M in A such that $p \in' M$. For the proof, apply Zorn's lemma to obtain a maximal ideal M that contains p'. Clearly $p \in' M$, for otherwise $1 = p \vee p' \in M$.

LEMMA 2. *The set X of all 2-valued homomorphisms on a Boolean algebra A is a closed subset of the Cantor space 2^A of all 2-valued functions on A.*

Proof. The definition of topology in 2^A implies that for each fixed p in A the value $x(p)$ depends continuously on the point x of 2^A. Since the set of points where two

continuous functions are equal is always a closed set, it
follows that $\{x : x(p') = (x(p))'\}$ is closed in 2^A for each p
in A. Forming the intersection of all these sets, we conclude that those 2-valued functions on A that preserve
complementation form a closed subset of 2^A. A similar
argument, involving sets such as $\{x : x(p \vee q) = x(p) \vee x(q)\}$,
justifies the same conclusion for the join-preserving
functions.

Lemma 2 implies that the set X of all 2-valued homomorphisms on a Boolean algebra A has the structure of a
Boolean space in a natural way; we shall call that Boolean
space the *dual space* of A.

The following assertion, known as the *Stone representation theorem*, is the most fundamental result about the
relation between Boolean algebras and Boolean spaces.

THEOREM 6. *The second dual of every Boolean algebra
A is isomorphic to A. More explicitly, if B is the dual
algebra of the dual space X of A, and if $f(p) = \{x \in X : x(p) = 1\}$
for each p in A, then f is an isomorphism from A onto B.*

Proof. Since $x(p)$ is continuous in X, it follows that
$f(p)$ is clopen for each p in A, and hence that f maps A into
B. The verification that f is a homomorphism is purely
mechanical. Thus, for example,

$$f(p \vee q) = \{x : x(p \vee q) = 1\} = \{x : x(p) \vee x(q) = 1\}$$

$$= \{x : x(p) = 1\} \cup \{x : x(q) = 1\} = f(p) \cup f(q).$$

If $f(p) = 0$, that is $\{x : x(p) = 1\} = \varnothing$, then Lemma 1 implies
that $p = 0$; this means that f is one-to-one. Since the range
of every Boolean homomorphism is a Boolean algebra, the

clopen sets of the form $\{x : x(p) = 1\}$ constitute a field. Since two distinct 2-valued homomorphisms on A must disagree on some element of A, the field is separating, and, consequently, Lemma 17.1 implies that f maps A onto B.

COROLLARY. *Every Boolean algebra is isomorphic to a field of sets.*

THEOREM 7. *The second dual of every Boolean space X is homeomorphic to X. More explicitly, if Y is the dual space of the dual algebra A of X, and if $\phi\,(x)$ is the 2-valued homomorphism that sends each element P of A onto 1 or 0 according as $x \in P$ or $x \in' P$, then ϕ is a homeomorphism from X onto Y.*

Proof. To prove that ϕ is continuous, it is sufficient to prove that the inverse image of every clopen subset of Y is clopen in X. The proof follows from the fact that every clopen subset of Y is of the form $\{y : y(P) = 1\}$, where $P \in A$ (see Theorem 6); indeed the inverse image of the indicated set is exactly P. We conclude also that the inverse image of a non-empty clopen set in Y is never empty; since the clopen sets form a base for Y, this implies that the range of the function ϕ is dense in Y. The continuity of ϕ and the density of its range together imply that ϕ maps X onto Y. Since the clopen sets separate points in X, distinct points of X determine distinct 2-valued homomorphisms on A, so that ϕ is one-to-one.

It is sometimes convenient to indicate the relation between Boolean algebras and Boolean spaces by some special terminology and notation. By a *pairing* of a Boolean algebra A and a Boolean space X we shall mean a function that associates with every pair (p, x), where $p \in A$ and $x \in X$, an element of 2 in a certain particular way. If the value of the function is denoted by $\langle p, x \rangle$, then the requirements on the function can be expressed as follows: (1) $\langle p, x \rangle$ is continuous in x, and, by suitable choice of p, every 2-valued continuous function on X has this form; (2) $\langle p, x \rangle$ determines a homomorphism in p, and, by suitable choice of x, every 2-valued homomorphism on A has this form. Here are two typical examples. (1) let X be the dual space of a Boolean algebra A; write $\langle p, x \rangle = x(p)$. (2) Let A be the dual algebra of a Boolean space X; write $\langle P, x \rangle = 1$ or 0 according as $x \in P$ or $x \in' P$. More generally, it should be clear by now that if A and X are paired, then A is isomorphic to the dual algebra of X and X is homeomorphic to the dual space of A.

Exercises

(1) Prove that every proper ideal in a Boolean algebra is included in some maximal ideal.

(2) A principal ideal in a Boolean algebra is not a sub-algebra, but it constitutes a Boolean algebra in a natural way (see §§9 and 11). Is that Boolean algebra necessarily isomorphic to a subalgebra of the whole algebra? (Hint: use 2-valued homomorphisms of the small algebra.)

(3) Is every Boolean space a subspace of a Cantor space?

(4) Is every complete Boolean algebra isomorphic to a complete field of sets?

(5) Is every Boolean algebra isomorphic to a subalgebra of a complete algebra?

§ 19. Duality for ideals

The topological duality theory of Boolean algebras, introduced in the preceding two sections, pervades and enriches the entire subject. Each of the two halves of the theory (algebras and spaces) suggests interesting questions about the other half. By means of the theory it is in principle possible to dualize every fact and every concept, converting algebraic facts and concepts into topological ones, and vice versa. In almost every case the dualization is worth while; it is often useful and illuminating, and, at the very least, it is amusing.

The following trivial example serves to illustrate the meaning of topological duality. Question: what can be said about the dual of a finite Boolean algebra? Answer: a Boolean algebra is finite if and only if it is the dual of discrete space. Reason: for compact Hausdorff spaces discreteness is the same as finiteness.

Finite Boolean algebras are atomic. A natural generalization of the problem of dualizing finiteness, and one that is somewhat less trivial, is the problem of dualizing atomicity. If, as before, X is a Boolean space and A is its dual algebra, then, by definition, an atom of A is a non-empty clopen subset of X that does not include any properly smaller non-empty clopen set. This implies that an atom of A is a singleton, namely the singleton of an isolated point of X. To say that A is atomic is to say that every clopen subset of X contains an isolated point. Since the clopen sets form a base, it follows that A is atomic if and only if the isolated points are dense in X. The other extreme has an equally satisfactory dual: A is non-atomic if and only if X is perfect.

The concept of countability has an interesting dual: the dual algebra A of a Boolean space X is countable if and only if X is metrizable. (Observe that for compact Hausdorff spaces metrizability is the same as the possession of a countable base.) Indeed, if A is countable, then the sets of A constitute a countable base for X. If, conversely, X has a countable base, then every base includes a countable subclass that is itself a base. This implies that there exists a countable base of clopen sets in X. The field generated by this base is still countable (see Exercise 8.5), and, by the corollary to Lemma 17.1; it coincides with A.

The duality theory for subsets of a Boolean algebra (for example, ideals and filters) is both more interesting and more useful than the duality theory for elements. The following definitions are the basic ones. If X is a Boolean space with dual algebra A, the dual of an ideal M in A is the union of the clopen sets belonging to M (equally correctly and more simply, the union of M), and the dual of an open subset U of X is the class of all clopen sets P included in U. The principal facts about this kind of set-duality can be summarized as follows:

LEMMA 1. *The dual of every ideal is an open set, and the dual of every open set is an ideal. The second dual of every ideal and of every open set is itself. Duality between ideals and open sets is a one-to-one correspondence that associates $\emptyset$ to the trivial ideal $\{0\}$ and X to the improper ideal A. If M and N are ideals with duals U and V, respectively, then the dual of $M \cap N$ is $U \cap V$, and a necessary sufficient condition that $M \subset N$ is that $U \subset V$. If, for each j in a certain index set, M_j is an ideal with dual*

U_j, then the union $\bigcup_j U_j$ is the dual of the ideal generated by $\bigcup_j M_j$.

The proofs of all the assertions of the lemma are immediate from the definitions.

It is easy to examine the duals of various special concepts in ideal theory. Thus, for instance, all the ideals of A are either trivial or improper if and only if all the open subsets of X are either $\emptyset$ or X. In other words, the unique simple algebra 2 is the dual of a singleton. The dual of a principal ideal is a clopen set, namely the generator. The dual of a maximal ideal is a maximal open set, that is, the complement of a singleton.

If M is an ideal in A, then $\{p : p' \in M\}$ is a filter in A; if U is an open set in X, then U' is a closed set in X. It follows that the duality between ideals and open sets induces a similar duality between filters and closed sets. The open duality is order-preserving; the closed duality is order-reversing. Thus, for example, the closed set corresponding to a maximal filter is a minimal closed set, that is, a singleton.

Exercises

(1) If "compact" is replaced by "locally compact" in the definition of Boolean spaces, most of the theory remains true. The dual of a locally compact but not compact Boolean space is a Boolean ring without a unit. A typical example of a non-compact Boolean space is obtained by omitting one point from a compact one. The act of restoring such an omitted point, that is, the one-point compactification, is the dual of the process of adjoining a unit (see Exercise 12.1). The dual of the empty Boolean space is the one-element (zero) Boolean ring (without a unit).

(2) The duality theory of ideals rests ultimately on the two relatively deep theorems of §18. This explains the fact that dualization can sometimes convert a non-trivial assertion into a complete triviality. For an example, dualize Exercise 18.1.

(3) The nowhere dense closed sets are of interest in a Boolean space, and so therefore are their complements, the dense open sets. Prove that the dual of a dense open set is a *dense ideal*, defined as follows. If M is a subset of a commutative ring R, the *annihilator* of M is the set of all elements p in R such that $pq = 0$ for all q in M. The annihilator of every set is an ideal. Motivated by the special case of Boolean rings and their topological duality theory, we call the set M dense if its annihilator is the trivial ideal.

(4) Prove that every countable Boolean algebra is isomorphic to a field of subsets of a countable set. (Hint: a compact metric space is separable.)

(5) What is the algebraic dual of separability? What about the first countability axiom (there is a countable base at each point)?

§20. Duality for homomorphisms

To establish a dual correspondence between structure-preserving mappings of Boolean algebras and Boolean spaces, it is best not to give preferential treatment to either. A good way to stay neutral is to use the concept of pairing introduced in §18. Suppose, accordingly, that A is a Boolean algebra and X is a Boolean space, and suppose that $\langle p, x \rangle$ represents all continuous 2-valued functions on X and all 2-valued homomorphisms on A. Suppose, moreover, that B

and Y are a similarly paired pair. The purpose of this section is to make a connection between continuous mappings
(from X into Y) and homomorphisms (from B into A). The
basic facts, on which subsequent definitions and theorems
depend, can be stated as follows:

THEOREM 8. *There is a one-to-one correspondence
between all continuous mappings ϕ from X into Y and all
homomorphisms f from B into A such that*

$$(1) \qquad \langle q, \phi(x) \rangle = \langle f(q), x \rangle$$

*identically for all q in B and all x in X. Each of ϕ and f is
called the dual of the other; the second dual of either one
is itself. The homomorphism f is one-to-one if and only if
ϕ maps X onto Y; the mapping ϕ is one-to-one if and only
if f maps B onto A.*

Proof. Fix ϕ and consider $\langle q, \phi(x) \rangle$. As a function
of q, for fixed x, it corresponds to (we shall just say: it is)
an element of Y, namely $\phi(x)$. This yields nothing new.
The novelty comes from considering $\langle q, \phi(x) \rangle$ as a
function of x for fixed q. Since it is the composite of the
two continuous functions $x \to \phi(x)$ and $y \to \langle q, y \rangle$, it
is a continuous 2-valued function on X. As such, it is given
by an (obviously unique) element p of A, so that

$$\langle q, \phi(x) \rangle = \langle p, x \rangle$$

identically in x. Denote the passage from q to p by f, that
is write $p = f(q)$. The proof that f is a Boolean homomorphism
is a mechanical computation. Here, for instance, is the
proof that f preserves complementation:

$$\langle f(q'), x \rangle = \langle q', \phi(x) \rangle = \langle q, \phi(x) \rangle' = \langle f(q), x \rangle' .$$

To make the definition of the dual homomorphism and the fact that it is a homomorphism more intuitive, suppose that A and B are the dual algebras of X and Y, respectively, and that the pairings are given by evaluating the characteristic function of the first coordinate at the second coordinate. In this case the fundamental duality equation (1) can be expressed as follows:

$$\phi (x) \in Q \text{ if and only if } x \in f(Q).$$

Since $\phi (x) \in Q$ if and only if $x \in \phi^{-1}(Q)$, this means that $f = \phi^{-1}$, or, more precisely, that f is the restriction of ϕ^{-1} to the class of clopen subsets of X.

Now fix f and consider $\langle f(q), x \rangle$. As a function of x, for fixed q, it is an element of A, namely $f(q)$. This yields nothing new. The novelty comes from considering $\langle f(q), x \rangle$ as a function of q for fixed x. Since it is the composite of the two homomorphisms $q \to f(q)$ and $p \to \langle p, x \rangle$, it is a 2-valued homomorphism on B. As such, it is given by an (obviously unique) element y of Y, so that

$$\langle f(q), x \rangle = \langle q, y \rangle$$

identically in q. Denote the passage from x to y by ϕ, that is, write $y = \phi (x)$. The definition of ϕ implies that

$$\phi^{-1}(\{y : \langle q, y \rangle = 1\}) = \{x : \langle f(q), x \rangle = 1\}.$$

Since every clopen subset of Y is given by some q in B, and since the clopen sets form a basis for Y, it follows that ϕ is continuous. To make the definition of ϕ more intuitive, suppose that X and Y are the dual spaces of A and B, respectively, and that the pairings are defined by evaluating the second coordinate at the first coordinate. In this case the fundamental duality equation (1) can be expressed as follows:

$$(\phi(x))(q) = x(f(q)).$$

The validity of this for all q says simply that $\phi(x)$ is equal to the composition $x \circ f$.

If f is the dual of ϕ and ψ is the dual of f, then, identically,

$$\langle q, \phi(x) \rangle = \langle f(q), x \rangle = \langle q, \psi(x) \rangle ,$$

and therefore $\phi = \psi$. If, finally, ϕ is the dual of f and g is the dual of ϕ, then, identically,

$$\langle f(q), x \rangle = \langle q, \phi(x) \rangle = \langle g(q), x \rangle ,$$

and therefore $f = g$.

It remains to prove the epi-mono assertions. For this purpose it is convenient to specialize; we shall assume that the algebra A is the dual of the space X. (The specialization is, of course, only notational; there is no real loss of generality here.) Consider now the following five assertions, each of which is easily seen to be equivalent to its neighbors: (1) ϕ maps X onto Y; (2) $Y - \phi(X) = \emptyset$; (3) every clopen subset of $Y - \phi(X)$ is empty; (4) if a clopen subset Q of Y is such that $\phi^{-1}(Q) = \emptyset$, then $Q = \emptyset$; (5) f is one-to-one. This proves the equivalence of (1) and (5). (To go from (2) to (3) recall that $Y - \phi(X)$ is always open.) Consider, finally, the following four assertions, each of which is equivalent to its neighbors: (1) ϕ is one-to-one; (2) the inverse images under ϕ of clopen subsets of Y separate points in X; (3) every clopen subset of X is the

inverse image under ϕ of some clopen subset of Y; (4) f
maps B onto A. (To go from (2) to (3) recall that the inverse
images under ϕ of clopen subsets of Y constitute a field of
subsets of X, and use Lemma 17.1.)

The proof of the fundamental duality theorem for homomorphisms is complete.

COROLLARY. *If ϕ is a continuous mapping from a
Boolean space X into a Boolean space Y, and if f is the
homomorphism dual to ϕ, then the dual of the kernel of f is
the complement of the range of ϕ.*

In loose language the corollary can be expressed as
follows: to divide an algebra by an ideal is the same as to
discard an open set from a space.

The epi-mono duality for structure-preserving maps implies a useful sub-quotient duality for the structures themselves. To see how this goes, suppose that the Boolean
algebras A and B are paired with the Boolean spaces X and
Y, respectively, and suppose that X is a subspace of Y.
This implies that there is a natural mapping ϕ (namely the
identity) from X into Y. Since ϕ is one-to-one, the dual
homomorphism f maps B onto A, so that A is isomorphic to
a quotient of B. (If, in fact, B is given as the dual of X,
then A is isomorphic to B/M, where the ideal M is the dual
of the open set $Y - \phi(X)$.) Inversely it is clear that every
quotient of B determines a subspace of Y. In the other
direction, suppose that B is a subalgebra of A. There is
then a natural homomorphism f (namely the identity) from B
into A. Since f is one-to-one, the dual mapping ϕ maps X
onto Y, so that Y is isomorphic to a quotient-space of X.
Inversely it is clear that every quotient of X determines a
subalgebra of A.

To see a non-trivial application of the duality theorem for homomorphisms, we consider the construction of free Boolean algebras (cf. § 10).

THEOREM 9. *For every set I, the dual algebra of the Cantor space 2^I is freely generated by a set of the same power as I.*

Proof. Write $Y = 2^I$, let B be the dual algebra of Y, and define a mapping h from I into B by $h(i) = \{y : y_i = 1\}$. Since h is one-to-one, the image $h(I)$ has the same power as I. The field generated by $h(I)$ is a base (by the definition of topology in Y) and therefore (by the corollary to Lemma 17.1) $h(I)$ generates B. We shall prove that B is free on $h(I)$. Suppose therefore that A is an arbitrary Boolean algebra and that g is an arbitrary mapping from I into A. Let X be the dual space of A and for each x in X write $\phi(x) = x \circ g$. Clearly $\phi(x) \in 2^I$ for each x in X, so that ϕ maps X into Y. Since

$$(2) \qquad \phi^{-1}(h(i)) = \{x : x \circ g \in h(i)\} = \{x : x(g(i)) = 1\},$$

and since the elements of $h(I)$ and their complements form a subbase for Y, it follows that ϕ is continuous. Let f be the dual homomorphism from B into A. This means (see (1)) that $\phi(x) \in h(i)$ if and only if $x(f(h(i))) = 1$. Since, by (2), $\phi(x) \in h(i)$ if and only if $x(g(i)) = 1$, it follows that

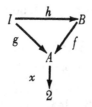

$x(f(h(i))) = x(g(i))$ for all x, and hence that $f \circ h = g$. In
other words, the homomorphism f is an extension of the
mapping $g \circ h^{-1}$, and the proof of the theorem is complete.

Exercises

(1) Prove that a finite Boolean algebra is free if and
only if the number of its atoms is a power of 2.

(2) Prove that every infinite free algebra is non-atomic.

(3) Prove that a countable non-atomic algebra is free.

(4) Prove that every Boolean algebra is isomorphic to a
quotient of a free one. (With Exercise 17.7, this gives a
solution of Exercise 14.3.)

(5) Give a topological solution of Exercise 18.2.

§ 21. Completion

By now we have seen the dual of every significant finite
algebraic concept that was introduced before; it is time to
turn to the infinite ones. What topological property, for
instance, characterizes a Boolean space whose dual algebra
is known to be complete? The answer is a weird but
interesting part of pathological topology.

A Boolean space is called *complete* if the closure of
every open set is open. (Observe that every compact
Hausdorff space with this property is automatically a Boolean space.) Complete Boolean spaces are sometimes
called *extremally disconnected* spaces. Completeness is a

self-dual property: a space is complete if and only if the interior of every closed set is closed. At first glance it is not at all obvious that non-trivial (that is non-discrete) complete spaces exist. It turns out, however, that they exist in profusion; there are as many of them as there are complete Boolean algebras.

The brunt of the major theorem in this direction is carried by an auxiliary result that has other applications also. It is in effect a topological characterization of the suprema that happen to be formable in a not necessarily complete Boolean algebra.

LEMMA 1. *If* $\{P_i\}$ *is a family of elements (clopen sets) in the dual algebra* A *of a Boolean space* X, *and if* $U = \bigcup_i P_i$, *then a necessary and sufficient condition that* $\{P_i\}$ *have a supremum in* A *is that* U^- *be open. If the condition is satisfied, then*

$$\bigvee_i P_i = U^-;$$

that is, the algebraic supremum is the closure of the set-theoretic union.

Proof. Assume first that $P = \bigvee_i P_i$. Since P is closed and includes each P_i, it follows that $U^- \subset P$. The set $P - U^-$ is open. If it is not empty, then it includes a non-empty clopen set Q, and then $P - Q$ is a clopen set including all the P_i's and properly included in P. Since this contradicts the definition of supremum (least upper bound), it follows that $U^- = P$, and hence that U^- is open.

If, conversely, U^- is open, then it is clopen, and, of course, it includes all the P_i's. If P is a clopen set that includes all the P_i's, then $U \subset P$, and therefore, since

P is closed, $U^- \subset P$. This implies that the family $\{P_i\}$ does have a supremum in A, namely U^-.

COROLLARY. *If a family of elements in the dual algebra of a Boolean space has a supremum, then that supremum differs from the set-theoretic union by a nowhere dense set.*

Proof. By Lemma 1 the difference in question is exactly the boundary of the union; apply Lemma 4.5.

THEOREM 10. *The dual algebra A of a Boolean space X is complete if and only if X is complete.*

Proof. If A is complete and if U is an open set in X, apply Lemma 1 to the family of all clopen subsets of U. If X is complete and if $\{P_i\}$ is a family of elements of A, apply Lemma 1 to the union $\bigcup_i P_i$.

The Stone representation theorem implies easily that every Boolean algebra can be embedded into a complete one. It is often important to know that Boolean algebras have completions in this sense, but the rudimentary completion obtained directly from the representation theorem is not good enough for most purposes.

The appropriate concept can be defined (somewhat pedantically) as follows. A *completion* of a Boolean algebra A is a complete Boolean algebra B together with an embedding h (that is, a monomorphism) from A into B, such that (1) if $\bigvee_i p_i = p$ in A, then $\bigvee_i h(p_i) = h(p)$ in B, and (2) the complete algebra generated by $h(A)$ in B is B itself. A completion (B, h) is *minimal* if it is smaller than every other completion in the following sense: corresponding to every completion (C, k), there exists a complete monomorphism f from B into C such that $f \circ h = k$.

It is not obvious that there are any completions at all, let alone minimal ones. We shall presently prove the necessary existence theorem. First, however, we dispose of uniqueness; the following result shows that the minimal completion of every Boolean algebra A is uniquely determined to within an isomorphism that preserves A.

LEMMA 2. *If (B, h) and (C, k) are minimal completions of A, then there exists an isomorphism between B and C that interchanges $h(A)$ and $k(A)$.*

Proof. The minimality of B and C implies the existence of complete monomorphisms f and g (from B to C and from C to B, respectively) such that $f \circ h = k$ and $g \circ k = h$. It follows, by substitution, that $f \circ g \circ k = k$ and $g \circ f \circ h = h$. If two complete homomorphisms agree on a complete generating set, then they are identical. This implies that f and g are each other's inverses.

The existence theorem produces the minimal completion of an algebra by combining two steps each of which separately is familiar by now. Given a Boolean algebra, use duality to associate it with a topological space, and then use some general topology to associate with that space the algebra of regular open sets; the result is, in a natural way, the minimal completion of the given algebra.

THEOREM 11. *If A is the dual algebra of a Boolean space X, if B is the algebra of regular open sets in X, and*

if h is the identity mapping from A into B, then (B, h) is a
minimal completion of A.

Proof. By Lemma 7.1, B is a complete Boolean algebra.
Clearly h is a one-to-one mapping from A to B. To verify
that h is a homomorphism we need merely to observe that
for clopen sets the Boolean operations of B reduce to the
ordinary set-theoretic operations. Suppose next that $\{P_i\}$ is
a family of elements in A that has a supremum, say P, in A.
Let Q be the supremum of $\{P_i\}$ in B. If $U = \bigcup_i P_i$, then P
is U^-, by Lemma 1, and Q is the interior of U^-, by Lemma
7.1. Since U^- is open, it follows that $P = Q$, so that suprema
formable in A stay the same in B. One more step is needed
to prove that B is a completion of A: we must show that
there is no complete algebra properly between A and B. For
this purpose, let U be an arbitrary element of B. Since U is
open, it is the union of the clopen sets it includes. The
supremum of that family of clopen sets in B is $U^{\perp\perp}$ (Lemma
7.1) and therefore U. This implies that U belongs to the
complete algebra generated by A in B, and therefore B is
indeed a completion of A.

It remains to prove minimality. Suppose that (C, k) is a
completion of A. If $U \in B$, write U as the union of all the
clopen sets it includes, $U = \bigcup_i P_i$, and define $f(U)$ as the
supremum of the family $\{k(P_i)\}$ in C. We shall show that f
is a complete monomorphism such that $f \circ h = k$. The last
part is the easiest. It says that if U happens to be clopen,
then $f(U) = k(U)$. The reasoning runs as follows. If U is
clopen, then U is not only the supremum but in fact the
largest element of $\{P_i\}$. This implies that $k(U)$ is the largest
element of $\{k(P_i)\}$, and hence that $k(U)$ is the supremum of
that family, as promised. Knowing that $f \circ h = k$ we can
conclude also that $f(U)$ can be 0 only when U is empty.
Indeed, if $f(U) = 0$, then $f(P) = 0$ for all clopen subsets P of

U, and therefore $k(P) = 0$ for all such sets. Since k is given as a monomorphism, it follows that every clopen subset of U is empty, and hence that U is empty. We know therefore that k is a monomorphism, provided it is a homomorphism at all. If, moreover, we borrow from the future the fact that f is a complete homomorphism, we can also conclude that f maps B onto C. Reason: the range of f is a complete sub-algebra of C that includes the range of k.

Let U be an element of B and write both U and its complement (that is, $U^\perp$) as unions of clopen sets; say $U = \bigcup_i P_i$, $U^\perp = \bigcup_j Q_j$. Since $P_i \cap Q_j = \varnothing$, it follows that $k(P_i) \cap k(Q_j) = 0$ for all i and j, and hence (see Lemma 7.4 and Exercise 7.8) that $f(U) \cap f(U^\perp) = 0$. The fact that $U \vee U^\perp = 1$ in B implies that there is no clopen set properly smaller than X that includes all the P's and Q's, so that $\bigvee_{i,j} (P_i \vee Q_j) = 1$ in A. Since (C, k) is a completion of A, it follows that $\bigvee_{i,j} (k(P_i) \vee k(Q_j)) = 1$ in C, and hence that $f(U) \vee f(U^\perp) = 1$. (This argument makes use of an easy special case of the associative law, Lemma 7.3). Conclusion: $f(U^\perp) = (f(U))'$.

The last thing to prove is that f preserves all suprema. Suppose that $\{U^j\}$ is a family of elements of B and write $U = \bigvee_j U^j$. Since f is obviously order-preserving, it is clear that $\bigvee_j f(U^j) \leq f(U)$. To prove the reverse inclusion, suppose that Q is a clopen subset of U, so that

$$Q = Q \cap U = \bigvee_j (Q \cap U^j) = \bigvee_j (Q \cap \bigvee_i P_i{}^j),$$

where, of course, $\{P_i{}^j\}$ is the family of all clopen subsets of U^j. It follows that $Q = \bigvee_j \bigvee_i (Q \cap P_i{}^j)$ in A, and hence that $k(Q) = \bigvee_j \bigvee_i (k(Q) \wedge k(P_i{}^j)) = \bigvee_j (k(Q) \wedge f(U^j)) = k(Q) \wedge \bigvee_j f(U^j)$.

This implies that

$$k(Q) \leqq \; \bigvee_j f(U^j)$$

for every clopen Q in U, and hence, by the definition of f, that $f(U) \leqq \bigvee_j f(U^j)$. The proof of the theorem is complete.

Exercises

(1) Every complete Boolean algebra is isomorphic to the regular open algebra of some compact Hausdorff space. (Hint: in every space clopen sets are regular open sets; in a complete Boolean space the converse is true.)

(2) If the regular open sets of a Boolean space constitute a field, does it follow that the space is complete?

(3) Show that if I is infinite and if X is the Cantor space 2^I, then $\mathcal{P}(X)$ is not a completion of the dual algebra of X.

(4) Let A be the Boolean algebra generated by the left half-closed intervals in $[0, 1]$, let B be the quotient of the algebra of Lebesgue measurable sets in $[0, 1]$ modulo the ideal of sets of measure zero, and let C be the quotient of the algebra of Borel sets in $[0, 1]$ modulo the ideal of meager sets. Prove that both B and C are in a natural sense completions of A.

(5) The minimal completion of an algebra A has the property that every element is the supremum of the elements of A that it dominates. Does any other completion have this property?

(6) Imitate the construction of real numbers by Dedekind cuts to construct a completion by cuts for every Boolean algebra. Is the completion so obtained minimal?

(7) Prove that the minimal completion of an atomic algebra is isomorphic to the field of all subsets of the set of atoms.

(8) Prove that the minimal completion of a non-atomic algebra is non-atomic.

(9) Does the minimal completion of an algebra satisfying the countable chain condition satisfy that condition also?

§ 22. Boolean σ-spaces

A *Baire set* in a Boolean space is a set belonging to the σ-field generated by the class of all clopen sets. Clearly every Baire set in a Boolean space is a Borel set; the converse is not true in general. A trivial way to manufacture open Baire sets is to form the union of a countable class of clopen sets. The converse is true but not trivial. The converse implies that every open Baire set is an F_σ (that is, the union of a countable class of closed sets), and, consequently, every closed Baire set is a G_δ (that is, the intersection of a countable class of open sets). We shall prove the main result about the structure of open Baire sets by proving first that every closed Baire set is a G_δ. Observe that in a metric space every closed set is a G_δ; in a general topological space this not so. The proof of the following auxiliary result uses the fact about metric spaces just mentioned; the trick is to construct a suitable metric space associated with each given closed Baire set.

LEMMA 1. *Every closed Baire set is a G_δ.*

Proof. Let F be a closed Baire set in the Boolean space X, and let $\{P_n\}$ be a sequence of clopen sets such that F

belongs to the σ-field generated by $\{P_n\}$ (see Exercise 13.7). Let p_n be the characteristic function of P_n and write

$$d(x, y) = \sum_{n=1}^{\infty} \frac{1}{2^n} |p_n(x) - p_n(y)|$$

for all x and y in X. The function d is a metric except perhaps for strict positiveness. If, in other words, two points x and y are defined to be equivalent, $x \equiv y$, in case $d(x, y) = 0$, then the equivalence classes may be more than singletons. (It is trivial that the relation so defined is an equivalence.) Let U be the set of all equivalence classes. There is a natural mapping T from X onto U; the value of $T(x)$ is the equivalence class of x, for each x in X. If $T(x_1) = T(x_2)$ and $T(y_1) = T(y_2)$, then

$$d(x_1, y_1) \leqq d(x_1, x_2) + d(x_2, y_2) + d(y_2, y_1) = d(x_2, y_2),$$

and, by symmetry, the reverse inequality is also true, so that $d(x_1, y_1) = d(x_2, y_2)$. This implies that writing

$$e(u, v) = d(x, y),$$

whenever $u = T(x)$ and $v = T(y)$, unambiguously defines a metric e on U. The inverse image (under T) of each open sphere in U is an open set in X, so that T is continuous. A set in X is the inverse image of some set in U if and only if it consists of (that is, is the union of) equivalence classes. The class of sets with this property is a σ-field. If $x \equiv y$, that is $d(x, y) = 0$, then $p_n(x) = p_n(y)$ for all n, so that x and y belong to the same P_n's; this implies that each P_n belongs to the σ-field of unions of equivalence classes. It follows from the definition of generated σ-field that F also belongs to that σ-field, and hence that $F = T^{-1}(V)$ for some subset V of U. Since $T(T^{-1}(V)) = V$, we infer that V

is compact and therefore closed. The set V is therefore a G_δ; the inverse images of a countable class of open sets whose intersection is V form a countable class of open sets whose intersection is F.

COROLLARY. *Every open Baire set in a Boolean space is the union of a countable class of clopen sets.*

Proof. By Lemma 1, every open Baire set is the union of a countable class of closed sets, say $G = \bigcup_n F_n$. Since the clopen sets form a base, each F_n is covered by the clopen sets included in G, and hence, by compactness, each F_n is covered by a finite number of such clopen sets.

We shall say that a Boolean space is a *σ-space* in case the closure of every open Baire set is open. The role of Boolean σ-spaces in the theory of σ-algebras is the same as the role of complete spaces in the theory of complete algebras.

THEOREM 12. *The dual algebra A of a Boolean space X is a σ-algebra if and only if X is a σ-space.*

Proof. (Compare Theorem 10.) If A is a σ-algebra and if U is an open Baire set in X, then (by the corollary above) U is the union of a countable class of clopen sets; since this class has a supremum in A, it follows (Lemma 21.1) that U^- is open. If X is a σ-space and if $\{P_n\}$ is a countable class of elements of A, then $\bigcup_n P_n$ is a Baire open set in X; since the closure of that set is open, it follows (Lemma 21.1) that $\{P_n\}$ has a supremum in A.

Exercises

(1) Let I be an uncountable discrete space; give examples of open sets that are not Baire sets in the one-point compactification of I and in the Cantor space 2^I.

(2) Prove that a set in a Boolean space is a Baire set if and only if it belongs to the σ-field generated by the class of all open F_σ's. (This condition can be and is used to define Baire sets in topological spaces more general than Boolean spaces.)

(3) Prove that in every topological space (see Exercise 2) with a countable base, every Borel set is a Baire set.

(4) Is Lemma 1 true in arbitrary compact spaces? (See Exercise 2.)

(5) Prove that the closure of a Baire set in a Boolean space need not be a Baire set. What if the space is a σ-space?

§ 23. The representation of σ-algebras

We know that every Boolean algebra is isomorphic to a field, whereas a complete Boolean algebra need not be isomorphic to a complete field (since, for instance, it need not be atomic). It is natural to ask the intermediate question: is every σ-algebra isomorphic to a σ-field? The answer is no. We shall see, in fact, that if A is a non-atomic σ-algebra satisfying the countable chain condition, then A cannot be isomorphic to a σ-field. For an example of such an algebra consider the regular open algebra of a Hausdorff space with no isolated points and with a countable base.

Alternatively, consider either the reduced Borel algebra or the reduced measure algebra of the unit interval.

To prove the negative result promised above, suppose that A is a non-atomic σ-algebra satisfying the countable chain condition. We shall make use of the fact (Corollary of Lemma 14.1) that A is complete. Assume now that A is isomorphic to a σ-field; we may as well assume that A is a σ-field of subsets of a set X. Select a point x of X and consider the class E of all those sets in A that contain x. Since A is complete, E has an infimum in A, say P; since A satisfies the countable chain condition, E has a countable subclass $\{P_n\}$ such that $P = \bigwedge_n P_n$ (see Lemma 14.1). The fact that A is a σ-field implies that $P = \bigcap_n P_n$; since each P_n contains x, it follows that $P \neq 0$. As a non-zero element of the non-atomic algebra A, the set P has a non-empty proper subset Q in A. Either Q or $P - Q$ contains x; we may assume that Q does. This means that $Q \in E$, and implies therefore that $P \subset Q$ (recall that $P = \bigwedge E$). This in turn implies that $P = Q$, and, since Q was supposed to be a proper subset of P, the contradiction has arrived.

If a class of Boolean algebras is not large enough to represent every algebra of a certain kind, the next best thing to hope is that the homomorphic images of the algebras of the class will suffice for the purpose. We have just seen that the class of σ-fields is not large enough to represent every σ-algebra; next we shall see that the class of homomorphic images of σ-fields (and, in fact, σ-homomorphic images) is quite large enough. The following result resembles Theorem 4 (p. 58) in many details, in both statement and proof. It is almost certain that the two results are special cases of a common generalization; it is far from certain whether the formulation and proof of such a generalization would yield any new information or save any time.

THEOREM 13. *Suppose that B is the σ-field of Baire sets and M is the σ-ideal of meager Baire sets in a Boolean σ-space X. Corresponding to each set S in B there exists a unique clopen set f(S) such that S + f(S) ∈ M. The mapping f is a σ-homomorphism from B onto the dual algebra A of X, with kernel M, so that A is isomorphic to B/M.*

Proof. We proceed as in the proof of Theorem 4 (p. 58). Consider the class of all those subsets of X that are congruent modulo M to a clopen set. This class is a σ-field that contains every clopen set, and, therefore, by the definition of generated σ-field, contains every Baire set. Since, by the Baire category theorem, two clopen sets can be congruent modulo M only if they are equal, the existence and uniqueness of $f(S)$ is proved for every Baire set S. A straightforward verification yields the remaining statements of the theorem.

COROLLARY. *Every σ-algebra is isomorphic to some σ-field modulo a σ-ideal.*

Proof. By Theorem 12 (p. 99) every σ-algebra is isomorphic to the dual algebra of some σ-space. This corollary is known as *Loomis's theorem.*

For σ-algebras, just as for plain Boolean algebras, the representation and duality theory yields an elegant proof of the existence and representation of free algebras.

THEOREM 14. *For every set I, there exists a free σ-algebra generated by I, and, in fact, that algebra is isomorphic to the σ-field of all Baire sets in the Cantor space 2^I.*

Proof. Write $Y = 2^I$, let B be the dual algebra of Y, and let B^* be the σ-field of Baire sets in Y. Define a mapping

h from I into B by $h(i) = \{y : y_i = 1\}$, let h^* be the identity mapping from B to B^*. The Boolean algebra generated by $h(I)$ is B; the σ-algebra generated by $h^*(h(I))$ is B^*. We are to prove that B^* is, in fact, the free σ-algebra generated by $h^*(h(I))$.

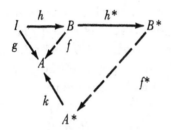

For this purpose we need to prove that every mapping g from I to an arbitrary σ-algebra A can be "extended" to a σ-homomorphism from B^* to A. We may and do assume that A is the dual algebra of a σ-space X. Let A^* be the σ-field of Baire sets in X; by Theorem 13 there is a σ-epimorphism, k say, from A^* to A.

Since (see Theorem 9, p. 89) B is free on $h(I)$, there exists a Boolean homomorphism f from B to A such that $f \circ h = g$. The homomorphism f is the dual of a continuous mapping ϕ from X into Y; this implies that $f(Q) = \phi^{-1}(Q)$ for every Q in B (see Theorem 8, p. 85). Let f^* be the σ-homomorphism from B^* into A^* defined by $f^*(S) = \phi^{-1}(S)$ for every S in B^*. The promised "extension" is the composition $k \circ f^*$.

Exercises

(1) Where does the proof of Theorem 13 make use of the assumption that X is a σ-space?

(2) Derive Loomis's theorem from Theorem 14.

(3) Is the generalization of Exercise 10.3 to free
σ-algebras true? (Compare Exercise 10.5.)

(4) Find an example of an m-algebra (for some infinite
cardinal m) that is not isomorphic to any m-field modulo an
m-ideal. (Hint: suppose that B is an m-field, M is an m-ideal
in B, and f is the projection of B onto B/M, where m is
greater than or equal to the power of the continuum. Prove
that if $I = \{1, 2, 3, \cdots\}$ and $J = 2$, then

$$\bigvee_{a \in 2^I} \bigwedge_{i \in I} (p_i + a(i)) = 1$$

for every sequence $\{p_i\}$ of elements of B/M. The idea is that
such a relation does hold in B and is preserved by f. The
regular open algebra of $(0, 1)$ does not satisfy this condition.)

§24. Boolean measure spaces

A *Boolean measure space* is a Boolean σ-space X together with a normalized measure on the σ-field of Borel
sets in X, such that non-empty open sets have positive
measure and nowhere dense Borel sets have measure zero.
The last condition is a very strange one. At first glance it
might seem that since a nowhere dense set is topologically
small and a set of measure zero is measure-theoretically
small, it is fitting and proper that the one should imply the
other. A little measure-theoretic experience (with Lebesgue
measure in Euclidean spaces, for instance) shows, however,
that the implication is not at all likely to hold. The results
of this section will show that Boolean measure spaces, in
which the implication is assumed to hold, have rather

pathological and almost paradoxical properties. The reason for considering them anyway is that measure algebras are important, and, as it turns out, Boolean measure spaces are exactly the duals of measure algebras.

We proceed to establish the notation that will be used in this section. Let X be a Boolean σ-space with dual algebra A (which is, therefore, a σ-algebra; see Theorem 12, p. 99). Let B be the σ-field of Borel sets in X, let M be the σ-ideal of meager Borel sets, and let f be the natural σ-epimorphism (Theorem 4, p. 58) from B onto the regular open algebra of X with kernel M.

LEMMA 1. *If ν is a normalized measure on B such that non-empty open sets have positive measure and nowhere dense Borel sets have measure zero, and if μ is the restriction of ν to A, then μ is a positive, normalized measure on A (so that A together with μ is a measure algebra).*

Proof. The only thing that needs proof is that μ is countably additive on A. Suppose that $\{P_n\}$ is a disjoint sequence of elements of A (clopen sets in X); write $U = \bigcup_{n=1}^{\infty} P_n$ and $P = \bigvee_{n=1}^{\infty} P_n$. By Lemma 21.1, $P = U^-$. Since, by Lemma 4.5, $U^- - U$ is nowhere dense, so that, by assumption, $\nu\,(U^- - U) = 0$, it follows that $\mu(P) = \nu(U)$. The countable additivity of μ on A is now an immediate consequence of the countable additivity of ν on B.

LEMMA 2. *If μ is a positive, normalized measure on A, then f maps B onto A. If $\nu\,(S) = \mu\,(f(S))$ for every S in B, then ν is a normalized measure on.B such that non-empty open sets have positive measure and such that the sets of measure zero are exactly the meager sets.*

Proof. The algebra A together with the measure μ is a measure algebra, and therefore complete (Corollary of

Lemma 15.3). It follows that the space X is complete (Theorem 10, p. 99), and hence that every regular open set in X is clopen (compare Exercise 21.1). This proves the first sentence of the lemma. The second sentence is an immediate consequence of Lemma 15.2.

COROLLARY. *The dual algebra A of a Boolean space X is a measure algebra if and only if X is a Boolean measure space.*

In the rest of the section we shall assume that X and A have not only the topological and algebraic properties originally required, but also the measure-theoretic structure (the measures ν and μ) described in Lemmas 1 and 2. This additional structure has profound and surprising effects on the topology of X. Thus, for instance, every open set is included in a clopen set of the same measure (namely its own closure). In other words, every open set is almost clopen; next we shall see that something like this is true for arbitrary Borel sets also.

Consider, indeed, those Borel sets whose measure can be approximated arbitrarily closely by clopen sets, from both inside and outside. More precisely, we shall say (temporarily) that a Borel set S is *regular* in case sup $\mu(P)$ = $\nu(S)$ (where the supremum is extended over all clopen sets P included in S) and inf $\mu(Q) = \nu(S)$ (where the infimum is extended over all clopen sets Q including S).

LEMMA 3. *Every Borel set is regular.*

Proof. We have already noted that an open set can be approximated arbitrarily closely by clopen sets from above. To approximate an open set U from below, consider the class of all clopen sets included in U, take a countable subclass with the same supremum, and take a finite sub-class whose union has very nearly the same measure. The self-dual character of the definition of regularity implies that the complement of a regular Borel set is regular, so that, in particular, every closed set is regular. If S_n is a nowhere dense Borel set, $n = 1, 2, 3, \cdots$, and therefore a set of measure zero, then the same is true of the closed set S_n^-. Since S_n^- is included in a clopen set of very small measure, it follows that every meager Borel set is included in an open set of small measure, and hence, by the already known facts for open sets, in a clopen set of small measure. By Theorem 4, p. 58 every Borel set S is congruent modulo meager sets to some clopen set P. If Q is a clopen set of small measure including the meager set $S + P$, then $P \cup Q$ is a clopen set that approximates S from above. Applying this result to S', we obtain the approximability of S by clopen sets from below, and the proof of the lemma is complete.

Lemma 3 says something very strong about the measure ν; the property it ascribes to ν is considerably stronger than the familiar measure-theoretic properties of regularity and completion regularity.

LEMMA 4. *Every Borel set has the same measure as its closure.*

Proof. If S is a Borel set, then, by Lemma 3, there exist clopen sets Q_n including S such that $\nu(Q_n - S) < 1/n$, $n = 1, 2, 3, \cdots$. The intersection of these clopen sets is a closed set that includes S and has the same measure as S.

LEMMA 5. *A Borel set of measure zero is nowhere dense.*

Proof. If $\nu(S) = 0$, then, by Lemma 4, $\nu(S^-) = 0$. This implies, by Lemma 3, that S^- includes no clopen set, and therefore also no open set.

LEMMA 6. *Every meager set is nowhere dense.*

Proof. If $S = \bigcup_n S_n$, where each S_n is nowhere dense, then each S_n^- is nowhere dense. It follows that $\nu(S_n^-) = 0$ (the reason for forming S_n^- is that S_n is not known to be measurable), and therefore S is included in the Borel set $\bigcup_n S_n^-$ of measure zero. The conclusion follows from Lemma 5.

Exercises

(1) Prove that in a Boolean measure space the boundary of every Borel set has measure zero.

(2) Prove that the dual space of the reduced measure algebra of [0, 1] is not separable. (Hint: use Lemma 4 and the fact that the algebra is non-atomic.)

(3) Use Exercise 2 to show that the reduced measure algebra and the reduced Borel algebra of [0, 1] are not isomorphic by showing that the dual space of the latter is separable. (Hint: For each t in [0, 1] define a proper ideal M_t in the regular open algebra A of [0, 1] thus: $U \in M_t$ if and only if $t \in U^\perp$. Since every proper ideal is included in some maximal ideal, there exists a 2-valued homomorphism x_t on A such that if $U \in M_t$, then $x_t(U) = 0$. If f is an isomorphism from the dual algebra of the dual space X of A to A itself, and if $t \in f(P)$, where P is a clopen subset of X, then $x_t \in P$. The set of x_t's with t rational is dense in X.)

§ 25. Incomplete algebras

The quotient of a Boolean algebra modulo an ideal may turn out to have a higher degree of completeness than one has a right to expect. Thus, for instance, the reduced Borel algebra and the reduced measure algebra of the unit interval are not only σ-algebras, which is all that the general theory can predict, but even complete. A few observations of this kind are likely to tip the balance of expectations too far over to the optimistic side. The purpose of this section is to provide a counterbalance in the form of some counter-examples. In other words, we shall obtain a few negative results: we shall see that certain quotient algebras are not complete.

The natural questions in this direction are obtained from the ones already answered by changing either the algebra or the ideal. The Borel sets modulo meager Borel sets in [0, 1] constitute a complete Boolean algebra; what about the Borel sets modulo countable sets, and what about all sets modulo meager sets? In deriving some of the answers we shall make use of the continuum hypothesis. This is sometimes avoidable; since, however, it simplifies and shortens the argument in any case, and especially since the purpose of the discussion is not to build the theory but merely to give warning of some danger spots, the effort of avoidance is not worth the trouble.

A typical result is that if X is an uncountable set, then the field B of all subsets of X modulo the ideal M of countable sets in X is not a complete Boolean algebra. To illustrate the argument, consider the special case in which X is the

Cartesian plane. The proof exhibits a concrete subset E of B/M that has no supremum. Let f be the projection from B to B/M; let E be the set of all those elements of B/M that have the form $f(S)$ for some vertical line S. The best way to prove that E has no supremum is to show that to every upper bound of E there corresponds a strictly smaller upper bound. Suppose, accordingly, that $f(S) \leqq p$ for all vertical lines S. Since f maps B onto B/M, there exists a subset P of X such that $p = f(P)$. To say $f(S) \leqq f(P)$ means that P contains all but countably many of the points of S. Since each S under consideration is uncountable, it follows that P contains at least one point in each S; let Q be a subset of P that contains exactly one point in each S. Clearly $f(S) \leqq f(P - Q)$ for each S; since, however, Q is uncountable (there are uncountably many vertical lines), it follows that $f(Q) \neq 0$ and hence that $f(P) - f(Q) \neq f(P)$.

The proof proves more than the statement states. Clearly the plane has nothing to do with the matter; any set in a one-to-one correspondence with the plane would do as well. The exact cardinality of X is also immaterial; all that matters is that X be uncountable. Indeed, since $m^2 = m$ for every infinite cardinal number m, there is always a one-to-one correspondence between X and X^2 (provided only that X is infinite), and the proof works again. Still another glance at the proof shows that the countability of the sets in the ideal M did not play a very great role; what mattered was that singletons belong to M and sets in one-to-one correspondence with X do not. On the basis of this last observation even the assumption that X is uncountable can be dropped; here is what remains.

LEMMA 1. *If B is the field of all subsets of an infinite set X, and if M is an ideal in B containing all singletons and not containing any set in one-to-one correspondence with X, then the algebra B/M is not complete.*

The lemma includes the statement we started with as a special case; as another special case it contains the statement that the algebra of all subsets of an infinite set modulo the ideal of finite sets is never complete, not even if the basic set is merely countable.

Lemma 1 is a relatively crude result, but its proof contains, in skeletal form, the two constructions that yield the more delicate results obtainable along these lines. The first step is to construct the set of vertical lines; in abstract terms, the problem is to construct a large disjoint class of sets none of which belongs to the prescribed ideal. The second step is to cut across the vertical lines; here the problem is to construct a large set whose intersection with each of the sets constructed before does belong to the ideal. The first of these constructions is the harder one; it is based on the following result of Ulam (*Fundamenta*, vol. 16).

LEMMA 2. *If X is the set of all ordinal numbers less than the first uncountable ordinal number Ω, then, corresponding to each natural number n and to each ordinal number a less than Ω, there exists a subset $S(n, a)$ in X such that the sets in each row of the array*

$$S(0, 0), \quad S(0, 1), \quad S(0, 2), \cdots, S(0, \omega), \cdots, S(0, a), \cdots$$

$$S(1, 0), \quad S(1, 1), \quad S(1, 2), \cdots, S(1, \omega), \cdots, S(1, a), \cdots$$

$$S(2, 0), \quad S(2, 1), \quad S(2, 2), \cdots, S(2, \omega), \cdots, S(2, a), \cdots$$

$$\vdots \qquad \vdots \qquad \vdots \qquad \vdots \qquad \vdots$$

$$S(n, 0), \quad S(n, 1), \quad S(n, 2), \cdots, S(n, \omega), \cdots, S(n, a), \cdots$$

$$\vdots \qquad \vdots \qquad \vdots \qquad \vdots \qquad \vdots$$

*are pairwise disjoint, and the union of the sets in each
column is cocountable.*

Proof. For each β in X select a sequence $\{k(a, \beta)\}$ of
type β (that is, $a < \beta$) whose terms are distinct natural
numbers. These sequences can be laid out in a triangular
array, as follows.

	0	1	2	$\cdots$	a	$\cdots$
0						
1	$k(0, 1),$					
2	$k(0, 2),$	$k(1, 2),$				
3	$k(0, 3),$	$k(1, 3),$	$k(2, 3)$			
.	.	.	.			
.	.	.	.			
.	.	.	.			
β	$k(0, \beta),$	$k(1, \beta),$	$k(2, \beta),$	$\cdots,$	$k(a, \beta),$	$\cdots$
.	.	.	.		.	
.	.	.	.		.	
.	.	.	.		.	

Let $S(n, a)$ be the set of all those β for which $k(a, \beta) = n$.
For example: to get $S(5, 2)$, consider the column labeled
2, and collect all those elements β for which the β entry
in that column has the value 5. If $\beta \in S(n, a)$, then $k(a, \beta) = n$;
since $k(a_1, \beta) \neq k(a_2, \beta)$ unless $a_1 = a_2$, it follows that,
for each n, the sets $S(n, a)$ are indeed pairwise disjoint.
If $a < \beta$, then

$$\beta \in S(k(a, \beta), a) \subset \bigcup_n S(n, a),$$

so that the union $\bigcup_n S(n, a)$ contains every β greater
than a; it follows that each such union is indeed cocountable.

To apply Lemma 2 we assume the continuum hypothesis.

COROLLARY 1. *The unit interval is the union of a disjoint class of power $\aleph_1$ consisting of sets none of which is meager.*

Proof. Establishing a one-to-one correspondence between [0, 1] and the set of all ordinal numbers less than Ω, we may and do assume that the sets $S(n, a)$ described in Lemma 2 are subsets of [0, 1]. Since each column of the square array of S's consists of countably many sets whose union is co-countable in [0, 1], it follows that at least one of the sets in each column must not be meager. Since there are uncountably many columns but only countably many rows, some row must contain uncountably many non-meager sets, and those sets, by Lemma 2, are pairwise disjoint. In case the union of the non-meager sets so obtained is not the entire interval, adjoin the complement of that union to one of them.

COROLLARY 2. *The unit interval is the union of a disjoint class of power $\aleph_1$ consisting of sets none of which has measure zero.*

Proof. Same as for Corollary 1; just interpret the word "meager" to mean "having measure zero".

We are now ready to imitate the argument (vertical lines) that lead to Lemma 1. This time let X be the unit interval, let B be the field of all subsets of X, and let M be the ideal of meager sets. By Corollary 1 there exists a disjoint family $\{S_i\}$ of power $\aleph_1$ consisting of sets not in M. Let f be the projection from B to B/M; let E be the set of all those elements of B/M that have the form $f(S_i)$ for some i. We shall show that the set E has no supremum by showing that to every upper bound p of E there corresponds a strictly smaller upper bound.

The preceding paragraph has the analogues of the vertical lines; the next problem is to cut across them. The technique here is based on the fact that every meager set is included in some meager F_σ. (Proof: every nowhere dense set is included in a closed nowhere dense set, namely its closure.) Since an easy argument shows that the cardinal number of the class of F_σ's is the power of the continuum, and since we have already assumed the continuum hypothesis, we may assume that all meager F_σ's occur as the terms of a family $\{R_i\}$ with the same set of indices as the family $\{S_i\}$.

Suppose now that P is a subset of X such that $f(P) = p$ (= the presumed upper bound of all $f(S_i)$). To say $f(S_i) \leq f(P)$ means that P includes all but a meager subset of S_i. Since R_i is meager but S_i is not, it follows that P contains at least one point in each $S_i - R_i$; let Q be a subset of P that contains exactly one point in each $S_i - R_i$. Clearly $f(S_i) \leq f(P - Q)$ for each i; since, however, Q is not included in any R_i, it follows that $f(Q) \neq 0$ and hence that $f(P) - f(Q) \neq f(P)$.

The proof is over; the time has come to see what it proves. The following statement (Sikorski) is a suitably general formulation of what the technique can be made to yield.

LEMMA 3. *Suppose that B is the field of all subsets of a set X, that M is an ideal containing all singletons, and that $\{R_i\}$ is a family of sets in M with the property that every set in M is included in some R_i. If there exists a disjoint family $\{S_i\}$, with the same set of indices, consisting of sets not in M, then the algebra B/M is not complete.*

A special case of the lemma, different from the one proved above, is that the algebra of all subsets of [0, 1] modulo the ideal of sets of measure zero is not complete. To deduce this conclusion from the lemma, let $\{R_i\}$ be the family of G_δ's of measure zero.

Exercises

(1) Prove Corollary 2 without assuming the continuum hypothesis.

(2) Let B be the field of all subsets of [0, 1] and let M be the ideal of countable sets. Is there a normalized measure on the algebra B/M?

§ 26. Products of algebras

A familiar way of making one new structure out of two old ones is to form their Cartesian product and, in case the structure involves some algebraic operations, to define the requisite operations coordinate-wise. Boolean algebras furnish an instance of this procedure. Since nothing is gained by restricting attention to two algebras at a time, we proceed at once to discuss arbitrary families. By the *product* of a family $\{A_i\}$ of Boolean algebras we shall understand their Cartesian product $\prod_i A_i$, construed as a Boolean algebra with respect to the coordinate-wise operations. This means that, for instance, 0 in $\prod_i A_i$ is defined by $0_i = 0$ for all i, and $p \vee q$ in $\prod_i A_i$ is defined by $(p \vee q)_i = p_i \vee q_i$ for all i. We shall indicate the product of finite or infinite sequences of Boolean algebras by such obvious and customary modifications of the symbolism as $\prod_{i=1}^n A_i$. For sequences of length two (and sometimes even for longer ones) we use the multiplication cross, so that $\prod_{i=1}^{2} A_i = A_1 \times A_2$. It is an immediate consequence of the definition that the product of a family of σ-algebras is a σ-algebra, and, similarly, the product of a family of complete algebras is complete.

In case each of a family of algebras $\{A_i\}$ is a field of subsets of a set X_i, then their product, say A, presents itself naturally as a field of subsets of the *disjoint union* of the given sets. One way to make the latter phrase precise is to form the set X of all ordered pairs (x, i) with $x \in X_i$. This set includes a copy (in an obvious sense) of each X_i, and these copies constitute a disjoint family of subsets of X. (The whole point of considering ordered pairs here is to force disjointness by means of the second coordinate.) To put the whole matter differently, suppose that a set X is the union of a disjoint family of non-empty subsets X_i. If A_i is a field of subsets of X_i, then A is naturally isomorphic to a field of subsets of X; the natural isomorphism is the one that assigns to an element P of A the subset $\bigcup_i P_i$ of X. (Recall that for each element P of $\prod_i A_i$, and for each index i, the coordinate P_i of P makes sense.)

If A is the product of a family $\{A_i\}$ of algebras, then, for each i, there is a natural epimorphism from A to A_i, namely the *projection* f_i defined by $f_i(p) = p_i$. If, moreover, B is an arbitrary Boolean algebra, and if, for each i, there is a homomorphism g_i from B to A_i, then there is a unique homomorphism g from B to A such that $f_i \circ g = g_i$ for all i. (Both existence and uniqueness are direct consequences of the definition of product.)

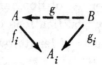

The algebra A and the family $\{f_i\}$ of homomorphisms are uniquely determined (to within isomorphism) by the property just ascribed to them. To prove this, suppose that B and $\{g_i\}$ have the same property. It follows that there exists a homomorphism f from A to B such that $g_i \circ f = f_i$ for all i; this, in turn, implies that $f_i \circ g \circ f = f_i$ and $g_i \circ f \circ g = g_i$ for all i. Since the role of $g \circ f$ in the first of these equations and the role of $f \circ g$ in the second is played by the identity automorphisms also (on A and on B, respectively), the assumed uniqueness proves that both f and g are isomorphisms, and that, in fact, each is the inverse of the other.

If $\{X_i\}$ is a family of Boolean spaces, we define their *sum* $\sum_i X_i$ as the dual space of the product of the corresponding algebras. Since the disjoint union of a finite family of Boolean spaces is a Boolean space in a natural way, it follows from the discussion of the product of fields of sets that the dual space of a finite product of algebras is (to within a homeomorphism) equal to the corresponding disjoint union of spaces. In other words, the concept of addition for Boolean spaces is an infinite generalization of the simple finite concept of disjoint union. Motivated by the additive terminology and notation, we shall use the plus sign for the finite concept, so that $\sum_{i=1}^{2} X_i = X_1 + X_2$.

If $A_i = 2$ for each element i of an infinite set I, then $\prod_i A_i$ is isomorphic to $\mathcal{P}(I)$. This shows that the sum of even the simplest spaces (singletons) can be something as unruly as the Stone-Cech compactification of an infinite discrete space (see Exercise 17.4).

The characterization of the product of a family of Boolean algebras by a family of homomorphisms can, of course, be dualized. The result is the following characterization of sums. If $\{X_i\}$ is a family of Boolean spaces, then there exists a Boolean space X, and, for each i, there exists a continuous one-to-one mapping ϕ_i from X_i into X, so that if Y is any Boolean space, and if, for each i, there exists a continuous mapping ψ_i from X_i to Y, then there exists a unique continuous mapping ψ from X to Y such that $\psi \circ \phi_i = \psi_i$ for all i.

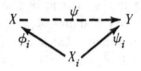

Exercises

(1) Suppose that $\{A_i\}$ is a family of Boolean algebras such that, for each i, there exists a positive normalized measure on A_i. Under what conditions does it follow that there exists a positive normalized measure on $\prod_i A_i$?

(2) Find Boolean algebras A, B, and C such that $A \times B = A \times C$ but $B \neq C$. (In other words, the cancellation law for products is false. Interpret the equal sign in this context to mean isomorphism.) Can the algebras be countable? Can they be finite?

(3) A product $\prod_i A_i$ includes two subalgebras, each of which might deserve some consideration as a kind of weak product of the family $\{A_i\}$. One subalgebra consists of those elements p for which $p_i \in 2$ for all but a finite set of indices i; the other, smaller, subalgebra consists of those elements p for which either $p_i = 0$ for all but a finite set of indices i or else $p_i = 1$ for all but a finite set of indices i. Give an example for which all three algebras are distinct. What can be said about the duals of these algebras?

§ 27. Sums of algebras

There are two ways to dualize arrow diagrams such as we met in the preceding section. What, for instance, does the diagram for products of algebras imply about the corresponding dual spaces? That is the first question; the answer is given by the diagram for sums of spaces. An equally natural question is this: What does the diagram for products of algebras become if the algebras and homomorphisms involved in it are replaced by spaces and continuous mappings? Two similar questions can be asked about the dualization of the diagram for sums of spaces. One of them leads back to products of algebras, and the other is the algebra dual of the space question just asked. The purpose of the present section is to answer the two as yet unanswered questions.

We proceed to the precise formulations. Suppose that $\{X_i\}$ is a family of Boolean spaces. Does there exist a Boolean space X, and does there exist, for each i, a continuous mapping ϕ_i from X onto X_i such that the requisite lifting condition is satisfied? The

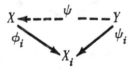

lifting condition says that if Y is a Boolean space and if, for each i, there exists a continuous mapping ψ_i from Y to X_i, then there exists a unique continuous mapping ψ from Y to X such that $\phi_i \circ \psi = \psi_i$ for all i. The answer is obviously yes; if X is the Cartesian product of the family $\{X_i\}$, with the product topology, and if the ϕ_i are the usual projections from a product space to its factors, then all the requirements are fulfilled. The only special fact that needs verification is that the clopen sets form a base for X (see Exercise 17.8). An argument similar to the one that proved the uniqueness of the product of Boolean algebras (to within an isomorphism) proves that there is a unique Boolean space (to within a homeomorphism) that, together with a suitable family of mappings, satisfies the lifting condition. It is natural to call the space we constructed the *product* of the given family of spaces and to use the multiplicative notation ($\prod_i X_i$, $X_1 \times X_2$, etc.) that this terminology suggests.

Suppose next that $\{A_i\}$ is a family of Boolean algebras. Does there exist a Boolean algebra A, and does there exist, for each i, a monomorphism f_i from A_i to A such that the transfer condition is satisfied? By the transfer condition we mean that if B is a Boolean algebra and if, for each i, there exists a homomorphism g_i from A_i to B, then there exists a unique homomorphism g from A to B such that $g \circ f_i = g_i$ for all i. The answer by now is obviously yes; just dualize the theory of products of Boolean spaces. To be more

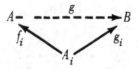

precise, let X_i be the dual space of A_i, let A be the dual algebra of $\prod_i X_i$, and, for each p in A_j, write $f_j(p)$ $= \{x \in \prod_i X_i : x_j(p) = 1\}$. We shall call the algebra the *sum* of the given family of algebras and we shall use the additive notation ($\sum_i A_i$, $A_1 + A_2$, etc.) that this terminology suggests. Standard arguments prove that the transfer condition uniquely determines the sum of a family of algebras, to within an isomorphism.

Sum and product constructions similar to the Boolean ones introduced above are useful for every known mathematical category, and, almost as a consequence of their universality, they are called by many different names. The terminology adopted above clashes head-on with some terms in common usage, but even so it is as nearly consistent with all already existing terminologies as any systematic usage could possibly be. No one will argue about products of spaces; that terminology is universally accepted. Products of algebras are almost as good (but not quite); the terminology is in harmony with accepted usage for groups, modules, and topological spaces. Instead of "product" a group-theorist would perhaps say "direct product", or, in the infinite case, "strong direct product", but that is close enough. Disagreements begin when group-theorists speak of "direct sum" or "strong direct sum". Even our "product" of Boolean algebras is sometimes called "direct sum", or, worse yet, "direct union". Our "sum" of Boolean spaces is not in common usage, but it does not seriously conflict with anything either; its sole competitor is "disjoint union", and that in the finite case only. The most radical departure is our "sum" of algebras. The word is in harmony with "weak direct sum" for modules, which, however, has also been called "weak direct product". The word is completely out of harmony with the usage in non-abelian group theory; the corresponding concept there is called "free product." Whether the word has the right intuitive connotations is perhaps arguable; at the very least a good case can be made out for it.

Exercises

(1) Show that a sum of complete algebras need not be complete. What about a finite sum? What about σ-algebras?

(2) If $A_i = 2$ for every element i of a set I, what is $\sum_i A_i$?

(3) Prove that if $A_i = 2 \times 2$ for every element i of a set I, then $\sum_i A_i$ is isomorphic to the free algebra generated by I.

§ 28. Isomorphisms of factors

A natural question about products of Boolean algebras (and of many other algebraic systems) is this: if each of two algebras is isomorphic to a factor of the other, does it follow that the two algebras are isomorphic. ("Factor" refers of course to the multiplication defined in §26.) The question can be reformulated and specialized in various interesting ways. For grammatical convenience we shall express the reformulations as statements rather than questions; the problem will then be to decide which statements are true and which ones false. For typographical convenience we shall use the sign of equality to denote isomorphism.

(1) If $D = A \times B$ and $A = D \times C$, then $A = D$.

(2) If $A = A \times B \times C$, then $A = A \times B$.

(3) If $A = A \times B \times B$, then $A = A \times B$.

(4) If $A = A \times 2 \times 2$, then $A = A \times 2$.

The assertions (1) and (2) are easily seen to be equivalent; (2) implies (3) (put $C = B$), and (3) implies (4) (put $B = 2$). Following Hanf, we shall settle the status of all these assertions by proving that (4) is false. The exposition is strongly influenced by several inspiring conversations with Dana Scott.

Let $\{a_n\}$ and $\{b_n\}$ be two countable sets, disjoint from each other, and let X be their union. A permutation T of X is defined by writing $T(a_n) = b_n$ and $T(b_n) = a_n$, $n = 1,2,3, \cdots$. The class of all those subsets of X that are invariant under T is a complete field of subsets of X; the atoms of that field are the couples $\{a_n, b_n\}$, $n = 1,2,3, \cdots$. Call a subset of X almost invariant if it differs from an invariant set by a finite set. The class A of all almost invariant sets is a field. The field A is atomic; its atoms are the singletons of X. Note that every infinite almost invariant set (that is, every infinite set in A) includes an infinite invariant set.

The algebra $A \times 2 \times 2$ can be described as follows. Adjoin two new points to X, say a_0 and b_0, and form all sets that differ from a set in A by a subset of $\{a_0, b_0\}$. The mapping S defined by $S(a_n) = a_{n+1}$ and $S(b_n) = b_{n+1}$, $n = 0,1,2, \cdots$, is a one-to-one correspondence between the enlarged set and the old one. The restriction of the inverse image map S^{-1} to sets in A is an isomorphism between A and $A \times 2 \times 2$.

The algebra $A \times 2$ can be described similarly. Adjoin one new point to X, say c, and form all sets that differ from a set in A by a subset of the singleton $\{c\}$. Observe that $A \times 2$ has an involution (an automorphism of period 2) that leaves exactly one atom fixed. (Extend T to a permutation U of $X \cup \{c\}$ by writing $U(c) = c$; the induced inverse image map is an involution of the sort described.) To prove that $A \neq A \times 2$, we shall show that A has no such involution.

Assume that, on the contrary, A has an involution U with exactly one fixed atom. We may assume, with no loss of generality, that that atom is one of the a's. The corresponding b is not left fixed by U; we may assume (typical special case) that its image is one of the a's. By applying this argument repeatedly we obtain an infinite sequence $\{a_{n_k}\}$ such that $U(a_{n_1}) = a_{n_1}$ and $U(b_{n_k}) = a_{n_{k+1}}$, $k = 1,2,3, \cdots$.

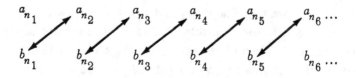

The set consisting of the a_{n_k}'s and the b_{n_k}'s with k congruent to 2 modulo 3 is invariant (under T) and therefore an element of A. The image of that set under U is infinite, but that image includes no non-empty subset invariant under T. This is a contradiction, proving that the assumption of the existence of U is untenable.

The phenomena so observed can be described in topological terms also. The dual space of A is just like the dual space of the algebra of all subsets of a countable set (which can alternatively be described as the Stone-Cech compatification of a countable discrete space), with one important modification: each isolated point is split into two distinct points, a red one and a blue one, say. The isolated points were dense before the split; they still are. Before the split interesting clopen sets were obtained by forming closures of infinite sets of isolated points; this is still true. What is different after the split is that two disjoint infinite sets of isolated points (for example, in case the given countable set consists of the positive integers, the red even numbers and the blue even numbers) can now have the same closure. The isomorphism facts proved above amount to this: adjoining (or discarding) two isolated points we get a homeomorphic space, but if we adjoin (or discard) only one isolated point, we do not.

There are several questions closely related to the ones we just answered. The counterexample to (4) is a large algebra (it has the power of the continuum); is there a countable one? If not, are there countable counterexamples to (3)? The answers are no (Exercise 28.7) and yes (§29) respectively. Is there a countable algebra A such that $A = A \times A \times A$ but $A \neq A \times A$? The answer is not known.

The corresponding questions for sums in place of products have not yet been attacked. It is not even known whether there exist Boolean algebras A and B such that $A + A = B + B$ but $A \neq B$:

(1) Prove that for the Boolean algebra A constructed above $A = A \times A$.

(2) Find two Boolean algebras A and D such that $A \times A = D \times D$ but $A \neq D$.

(3) Find a Boolean algebra D such that $D = D \times D \times D$ but $D \neq D \times D$.

(4) Find a Boolean algebra A such that $A = A \times 2 \times 2 \times 2$ but $A \neq A \times 2$ and $A \neq A \times 2 \times 2$.

(5) Find Boolean algebras A_1 and A_2 such that $A_1 \times A_2 = A_1 \times A_2 \times 2$ but $A_1 \neq A_1 \times 2$ and $A_2 \neq A_2 \times 2$.

(6) Prove that if A is a countable Boolean algebra with infinitely many atoms, and if B is a finite Boolean algebra, then $A = A \times B$. (Hint: dualize.)

(7) Prove that if A is a countable Boolean algebra, and if B and C are finite Boolean algebras such that $A = A \times B \times C$, then $A = A \times B$.

§29. Isomorphisms of countable factors

The purpose of this section is to show (following Hanf, as simplified, orally, by Dana Scott) that there exist countable Boolean algebras A and B such that $A = A \times B \times B$ but $A \neq A \times B$. The method of attack is topological; in fact, we shall construct Boolean spaces X and Y, each with a countable base, so that $X = X + Y + Y$ but $X \neq X + Y$. (The equal sign denotes homeomorphism here.)

We begin by constructing for each integer n $(= 0,1,2, \cdots)$ a Boolean space U_n, with countable base, and a distinguished point u_n of U_n, such that no neighborhood of u_n is homeomorphic to any neighborhood of any other point in any U_m (not even in U_n itself). Here is one way to do this: let U_n consist of a sequence of type ω^n in $[-1,0]$ converging to 0, together with the Cantor set in $[0, 1]$. The point 0 of U_n is then such that the derivatives of order less than n of every neighborhood of it contain isolated points, whereas the n-th derivative is perfect. No other point in any of the spaces under consideration can make that claim.

Next we form the union of a disjoint class consisting of exactly one copy of each of the spaces U_k with $k \geq n$; let Y_n be the one-point compactification, by y, of that union. Schematically Y_n may be represented in the form

$$n, \ n + 1, \ n + 2, \ \cdots \longrightarrow y,$$

where, for the sake of brevity, we have used the symbol for the integer n to denote the space U_n. We form also the union of a disjoint class consisting of exactly two copies of each of the spaces U_k with $k \geqq n$; let Z_n be the one-point compactification, by z, of that union. Schematically Z_n may be represented in the form

$$n, \ n + 1, \ n + 2, \ \cdots \longrightarrow z \longleftarrow \cdots, \ n + 2, \ n + 1, \ n.$$

We go on to form the union of a countable disjoint class consisting of copies of Z_0, and compactify it by one point $z*$. The result is represented schematically by the part of the subjoined diagram that lies above the unbroken dividing line. The part of the diagram below that line is a schematic representation of the union of a disjoint class consisting of exactly one copy of each Z_n and of exactly two copies of each Y_n, compactified by one point $y*$. Let X be the disjoint union of the two grand unions formed before, so that the whole diagram represents X. Clearly X is a Boolean space with a countable base.

Each copy of each u_n in X has a neighborhood that contains no other copy of that u_n or of any other. The u_n's are the only points of X with this property.

Every neighborhood of each copy of y in X contains a copy of almost all the u's (that is, all but a finite number), and some neighborhood of each y contains exactly one copy of each u. The y's are the only points in X with this property.

Every neighborhood of each copy of z in X contains at least two copies of almost all the u's, and some neighborhood of each z contains exactly two copies of each u. The z's are the only points in X with this property.

Every neighborhood of y^* contains almost all y's and almost all z's. The point y^* is the only point in X with this property.

Every neighborhood of z^* contains almost all z's, and some neighborhood of z^* contains no y's. The point z^* is the only point in X with this property.

The preceding paragraphs imply that if T is a homeomorphism of X onto X, then y^* and z^* are invariant under T, the set U_n^* of all u_n's is invariant under T for each n, the set Y^* of all y's is invariant under T, and the set Z^* of all z's is invariant under T.

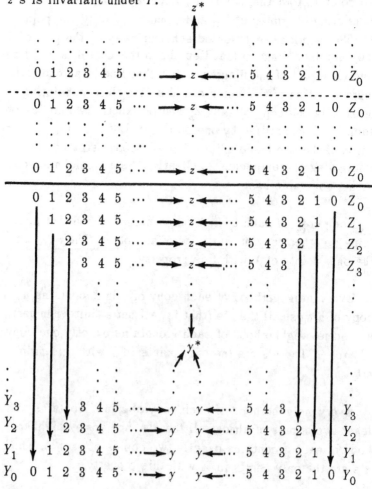

Let the space Y be the Y_0 already defined above. We are to prove that two copies of Y can be adjoined to X with impunity; we shall prove the equivalent assertion that two copies of Y can be discarded from X with impunity. Suppose, indeed, that the bottom row of the diagram is erased. To reconstruct the space X, take the 0's from the lowest Z_0 and give them to Y_1, take the 1's from Z_1 and give them to Y_2, and so on, as indicated by the long vertical arrows in the diagram; leave all other parts of X alone. The transformation so defined is a homeomorphism from the deleted space to the original X. The verification of this assertion is routine. The only excitement can come from a sequence chosen from the moving parts and converging to y^*; the construction guarantees that the transform of such a sequence still converges to the (fixed) point y^*.

The next and last thing to prove is that the adjunction of one copy of Y to X makes a difference; we shall prove the equivalent assertion that if X is diminished by discarding one copy of Y, say the right half of the bottom line, then the resulting space $\widetilde{X}$ is topologically distinguishable from X. Indeed: X has an involution (a self-homeomorphism of period 2) that leaves fixed each point of $Y^* \cup Z^* \cup \{y^*\} \cup \{z^*\}$, and nothing else. (Reflect the diagram about the central vertical axis.) We shall prove that $\widetilde{X}$ has no such involution. Suppose that, on the contrary, T is an involution whose set of fixed points is exactly $Y^* \cup Z^* \cup \{y^*\} \cup \{z^*\}$; our remaining task is to derive a contradiction from this supposition.

Let V be the part of X represented by the part of the diagram below the unbroken horizontal line. Since V is a compact subset of $\widetilde{X} - \{z^*\}$, the same is true of $T(V)$. This implies that there exists a dotted horizontal line (as indicated) such that $T(V)$ is below it. (The linguistic identification of parts of the diagram with corresponding

parts of the space $\tilde{X}$ is obvious and harmless.) Below the
dotted line there are an odd number of 0's. Since T maps
U_0^* into itself, and since no copy of u_0 is fixed under T, one
of those 0's (or, to be a little more precise, the copy of u_0
belonging to one of those 0's) is mapped above the dotted
line. The 0 (or 0's) to which this happens cannot be in V
(since $T(V)$ is below the dotted line). Conclusion: one of
the 0's between the two horizontal lines gets mapped above
the dotted line. What was just argued about the 0's is just
as true about the 1's, the 2's, etc. Since there are only a
finite number of rows between the two horizontal lines, it
follows that there is at least one such row with the property
that infinitely many of its parts get mapped above the dotted
line. Since from those parts a sequence of points converging
to some z (between the lines) can be selected, the continuity
of T implies that T moves some z from between the lines to
above the dotted line. The contradiction has arrived: the z's
must be fixed under T.

§ 30. Retracts

A Boolean algebra B is a *retract* of a Boolean algebra A
if there exist homomorphisms f and g mapping A and B into
B and A, respectively, such that $f \circ g$ is the identity mapping
on B. The condition implies that f is an epimorphism and g
is a monomorphism, so that a retract of A may be simultaneously regarded as a quotient algebra and a subalgebra of A.

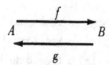

We shall meet quite a few arrow diagrams in what follows;
some conventions will be useful. An epimorphism will be

indicated by double-headed arrows ⟶⟶ ; monomorphism will be indicated by double-footed ones ⊩⟶ .
The adjoined diagram is, accordingly, a better representation of the definition of a retract than the one above. The

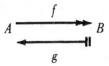

diagram obtained from a given one by reversing all arrows
and interchanging double-headed and double-footed ones is,
in a certain informal sense, the dual of that given one. In
this sense the concept of retraction (or, rather, its diagram)
is self-dual.

We shall say that an algebra B is an *absolute subretract*
if, roughly speaking, B is a retract of every algebra that
includes it. More precisely, B is an absolute subretract in
case corresponding to every monomorphism g from B to any
Boolean algebra A there exists an epimorphism f from A to
B such that $f \circ g$ is the identity on B. Dually, B is an
absolute quotient retract if B is a retract of every algebra
that maps onto it. In precise terms the requirement is that
to every epimorphism f from an arbitrary Boolean algebra A
to B there corresponds a monomorphism g from B to A such
that, again, $f \circ g$ is the identity on B.

The dual definitions make sense and are worth while in
the study of Boolean spaces. A Boolean space Y is a retract
of a Boolean space X

if there exist continuous mappings ϕ and ψ, as indicated in
the diagram, such that $\psi \circ \phi$ is the identity on X. The
definitions of the absolute concepts should be obvious by
now. Note that if $\langle A, X \rangle$ and $\langle B, Y \rangle$ are dual pairs
(see §18), then a necessary and sufficient condition that Y
be a retract of X is that B be a retract of A; in this sense
of duality absolute subretracts and absolute quotient retracts
are the duals of one another.

The rather natural definitions above are special cases of
some others that, on first glance, may look somewhat
artifical. Since, however, it turns out that the generalizations
have a very satisfying and useful theory, we proceed to
introduce them. We say, accordingly, that a Boolean algebra
B is *projective* if every homomorphism from B to a quotient
can be lifted to the numerator. More precisely, B is projective in case for every epimorphism f from an arbitrary Boolean
algebra A

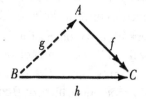

to an arbitrary Boolean algebra C, and for every homomorphism h from B into C, there exists a homomorphism g from
B into A such that $f \circ g = h$. Clearly every projective algebra
is an absolute quotient retract. (Take $C = B$ and let h be the
identity.) Similarly, B is *injective* if every homomorphism
from some subalgebra into B can be extended to the whole
algebra. More precisely, B is injective in case for every
monomorphism g from an arbitrary Boolean algebra C to an
arbitrary Boolean algebra A, and for every homomorphism h
from C into B, there exists a homomorphism f from A into B

such that $f \circ g = h$. Clearly every injective algebra is an absolute subretract. (Take $C = B$ and let h be the identity.)

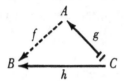

The dual definitions apply, of course, to Boolean spaces. A Boolean space Y is *projective* if every continuous mapping from Y to a quotient space can be lifted to the numerator; it is *injective* if every continuous mapping from some sub-space into Y can be extended to the whole space. The precise formulations, as well as the informal ones just given, are the topological duals of the corresponding algebraic definitions.

All the concepts defined in this section have a universal (and hence rather shallow) character; they apply with only minor modifications to modules, or groups, or topological spaces, and, in fact, to every known category of mathematical objects. The interested reader may pursue this comment for himself. We shall not even pause to give the appropriate examples and counterexamples in the cases of central inter-est (that is, Boolean algebras and Boolean spaces). We stay, instead, on or near the universal level, by deriving some elementary consequences of the definitions; the juicy existence and characterization theorems follow in later sections.

LEMMA 1. *Every retract of a projective algebra is pro-jective; every retract of an injective algebra is injective.*

Proof. Assume that B is a retract of a projective algebra $\tilde{B}$, with associated epimorphism k and monomorphism j. It is to be proved that for given A, C, f, and h, as in the diagram,

g can be contructed. Write $\hat{h} = h \circ k$, and, using the projectivity of $\tilde{B}$, lift $\hat{h}$ to $\tilde{g}$. The desired homomorphism is defined by $g = \tilde{g} \circ j$. The dual assertion for injective algebras ("dual" in the arrow diagram sense) is proved by a dual proof.

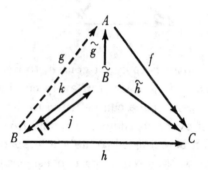

COROLLARY. *Every retract of a projective Boolean space is projective; every retract of an injective space is injective.*

Proof. Topological duality, from Lemma 1.

LEMMA 2. *If Y is the sum (disjoint union) of a finite family $\{Y_i\}$ of Boolean spaces, then each Y_i is a retract of Y; if, in fact, δ_i is the natural mapping (embedding) from Y_i to Y, then there exists a continuous mapping γ_i from Y onto Y_i such that $\gamma_i \circ \delta_i$ is the identity on Y_i.*

Proof. (Compare Exercises 18.2 and 20.5.) We may and do identify each Y_i with a clopen subset of Y. Map Y onto Y_i by mapping Y_i onto itself identically and sending every other point in Y onto an arbitrary but fixed point of Y_i.

LEMMA 3. *If Y is the product of a family $\{Y_i\}$ of Boolean spaces, then each Y_i is a retract of Y; if, in fact, γ_i is the projection from Y onto Y_i, then there exists a continuous mapping δ_i (which is necessarily one-to-one) from Y_i into Y such that $\gamma_i \circ \delta_i$ is the identity on Y_i.*

Proof. Map Y_i into Y by selecting an arbitrary but fixed point from each factor, except Y_i itself, and sending every point in Y_i onto that point of Y whose i coordinate is the given one and whose other coordinates are the selected points.

THEOREM 15. *The sum of a finite family of Boolean spaces is injective if and only if each one of them is injective.*

Proof. The "only if" follows from the Corollary of Lemma 1 and Lemma 2. To prove the converse, let Y be the sum of the finite family $\{Y_i\}$, let δ_i be the natural mapping from Y_i to Y, and

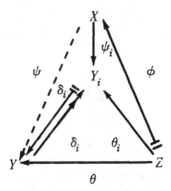

let γ_i be a continuous mapping from Y onto Y_i such that $\gamma_i \circ \delta_i$ is the identity on Y_i (Lemma 2). Suppose now that each Y_i is injective. It is to be proved that for given X, Z, ϕ, and θ, as in the diagram, ψ can be contructed. Write $Z_i = \theta^{-1}(\delta_i(Y_i))$, so that $\{Z_i\}$ is a finite disjoint family of clopen sets in Z. It follows that $\{\phi(Z_i)\}$ is a finite disjoint family of closed sets in X, and, consequently, there exists a disjoint family $\{X_i\}$ of clopen sets in X such that $\phi(Z_i) \subset X_i$ for each i. We may and do assume that $\cup_i X_i = X$. Let θ_i be the restriction of $\gamma_i \circ \theta$ to Z_i, and using the injectivity of Y_i extend θ_i to mapping ψ_i from X_i to Y_i. The desired mapping is defined by writing $\psi(x) = \delta_i(\psi_i(x))$ whenever $x \in X_i$.

COROLLARY. *The product of a finite family of Boolean algebras is projective if and only if each one of them is projective.*

THEOREM 16. *The product of a family of Boolean spaces is injective if and only if each one of them is injective.*

Proof. The "only if" follows from the Corollary of Lemma 1 and Lemma 3. To prove the converse, let Y be the product of the family $\{Y_i\}$, and let γ_i be the projection from Y into Y_i. Suppose now that each Y_i is injective. It is to be proved that for given X, Z, ϕ, and θ, as in the diagram, ψ can be constructed.

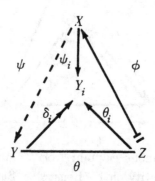

Write $\theta_i = \gamma_i \circ \theta$, and, using the injectivity of Y_i, extend θ_i to a mapping ψ_i from X to Y_i. The desired mapping is ψ, uniquely determined by $\gamma_i(\psi(x)) = \psi_i(x)$ for each i.

COROLLARY. *The sum of a family of Boolean algebras is projective if and only if each one of them is projective.*

§ 31. Projective algebras

We still have no examples of projective algebras; the following result provides infinitely many.

THEOREM 17. *Every free Boolean algebra is projective.*

Proof. Suppose that B is free on a subset I and let j be the identity mapping of I into B. It is to be proved that given A, C, f, and h, as in the diagram, g can be constructed. For every i

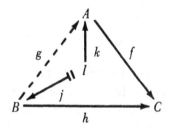

in I there exists an element $k(i)$ in A such that $f(k(i)) = h(j(i))$; the reason is that f is an epimorphism. Since B is free on I, there exists a unique homomorphism g from B to A such that $g \circ j = k$. Since $f \circ g$ agrees with h on I, the fact that I generates B implies the desired result.

COROLLARY 1. *Every Cantor space is injective.*

COROLLARY 2. *A Boolean algebra is projective if and only if it is a retract of a free algebra.*

Proof. A retract of a free algebra is a retract of a projective algebra (Theorem 17) and therefore projective (Lemma 30.1).

A projective algebra is an absolute quotient retract (§30); since every Boolean algebra is a quotient of a free one, it follows that a projective algebra is a retract of a free one.

COROLLARY 3. *A Boolean space is injective if and only if it is a retract of a Cantor space.*

COROLLARY 4. *A Boolean algebra is projective if and only if it is an absolute quotient retract.*

Proof. The "only if" was proved in §30. If an algebra is an absolute quotient retract, then, in particular, it is a retract of a free algebra and hence (Corollary 2) projective.

COROLLARY 5. *A Boolean space is injective if and only if it is an absolute subretract.*

Freedom is a rather severe structural restriction on a Boolean algebra and it is not too surprising that freedom implies projectivity. It is considerably more surprising that a cardinal number restriction can also imply projectivity; we proceed to prove an assertion of this kind, first in dual form.

LEMMA 1. *If I is a countable set, then every non-empty closed subset of the Cantor space 2^I is a retract of 2^I.*

Proof. We may and do assume that $I = \{1, 2, 3, \cdots \}$; all that we omit thereby is the trivial finite case. If x and y are in 2^I, write $d(x, y) = \sum_{i=1}^{\infty} \frac{1}{10^i} |x_i - y_i|$; the function d is a metric that induces the topology of 2^I. The uniqueness of the decimal expansions in which only 0's and 1's occur (no 9's) implies that if $d(x, y) = d(x, z)$, then $y = z$. It follows that if F is a non-empty closed subset of 2^I, then a transformation ϕ of 2^I into itself is unambiguously determined by writing $y = \phi(x)$ in case y is the point of F nearest to x. Straightforward verification proves that ϕ is continuous. The composition of ϕ with the natural embedding of F into 2^I is the identity mapping on F; this completes the proof.

COROLLARY 1. *Every Boolean space with a countable base is injective.*

Proof. Every such space is homeomorphic to a subset of a Cantor space with a countable base.

COROLLARY 2. *Every countable Boolean algebra is projective.*

We began this section with not enough examples of projective algebras; we have reached the point where it could seem that every algebra is projective. This is not so; the characterization in Corollary 2 of Theorem 17 can be used to obtain negative results as well as the positive ones obtained above. Indeed: every free algebra satisfies the countable chain condition, and, therefore, so does every subalgebra of a free algebra. Hence, in view of Corollary 2 of Theorem 17, to get an example of an algebra that is not projective, it is sufficient to exhibit an algebra that does not satisfy the countable chain condition. The finite-cofinite algebra of an uncountable set will do.

No satisfactory characterization of projective Boolean algebras (or, equivalently, of injective Boolean spaces) is known. Combining the results of this section and the preceding one we see that the product of a finite number of algebras each of which is the sum of an arbitrary number of countable algebras is projective; for all that is known now every projective algebra is included in this description.

Exercises

(1) Prove that every projective Boolean algebra has a finitely additive, positive, normalized measure whose values are dyadic rational numbers (that is, rational numbers whose denominator is a power of 2). (Hint: use Corollary 3 of Theorem 17 and the theory of Haar measure.)

(2) Prove that every infinite complete algebra has a subalgebra that is isomorphic to the field of all subsets of a countable set.

(3) Prove that an infinite complete algebra is never projective. (Hint: in view of Exercises 1 and 2 it is sufficient to prove that the field of all subsets of a countable set has no measure of the kind described in Exercise 1. Prove first that such a measure could not be countably additive on any infinite set. Reason: since it takes arbitrarily small values, it would then take irrational values. Hence: if the sum of the measures of the singletons in a set is subtracted from the measure of the set itself, the result vanishes just in case the set is finite. This implies the existence of a finitely additive, positive, normalized measure on the quotient of the field of all subsets of a countable set modulo the ideal of finite sets. Since, however, that quotient does not satisfy the countable chain condition, the result is a contradiction. This proof is due to D. S. Scott and H. F. Trotter.)

§ 32. Injective algebras

The theory of injective algebras is somehwat harder than the theory of projective algebras, but the extra difficulty buys considerably more satisfying information. Injectivity for Boolean algebras turns out to be closely connected with completeness; we begin with an auxiliary result on complete algebras.

LEMMA 1. *Every retract of a complete algebra is complete.*

Proof. Let f be an epimorphism from a complete algebra A to the algebra B, say, and let g be a monomorphism from B to A such that $f \circ g$ is the identity on B. It is to be proved that every family $\{q_i\}$ in B has a supremum in B. Write $p_i = g(q_i)$ and $p = \bigvee_i p_i$ (in A). Assertion: if $q = f(p)$, then $q = \bigvee_i q_i$ in B. Since $p_i \leq p$, it follows that $f(p_i) \leq q$ and hence that $q_i \leq q$ for all i (recall that $f(g(q_i)) = q_i$). If $q_i \leq r$ for all i, then $p_i \leq g(r)$ and therefore $p \leq g(r)$; this implies that $q \leq r$.

A part of the connection between injectivity and completeness becomes visible now.

THEOREM 18. *Every injective algebra is complete.*

Proof. We know that every algebra can be embedded into a complete one (§21). Since every injective algebra is an absolute subretract (§30), it follows that every injective algebra is a retract of a complete one; the conclusion now follows from Lemma 1.

The crucial result along these lines is the converse.

THEOREM 19. *Every complete algebra is injective.*

The proof depends on a lemma that has other applications; we shall refer to it as the *extension lemma*.

LEMMA 2. *Suppose that a Boolean algebra A is generated by a subalgebra C and an element r, and suppose that h is a homomorphism from C to B, say. Suppose also that p_* and p^* are elements of B such that $h(s) \leq p_*$ whenever $s \in C$ and*

$s \leqq r$, and $p^* \leqq h(t)$ whenever $t \in C$ and $r \leqq t$. *Under these conditions, to each element p of B, with $p_* \leqq p \leqq p^*$, there corresponds a unique extension f of h to a B-valued homomorphism on A such that $f(r) = p$.*

Proof. Since C and r generate A, every element of A has a (not necessarily unique) representation in the form $(s \wedge r) \vee (t \wedge r')$ with s and t in C. A straightforward calculation shows that the mapping f that sends each such element onto

$$(h(s) \wedge p) \vee (h(t) \wedge p')$$

is unambiguously determined and does everything that is required of it.

We are now ready to prove Theorem 19. Suppose, accordingly, that B is a complete Boolean algebra, that C is a subalgebra of a Boolean algebra A (with embedding g) and that h is a homomorphism from

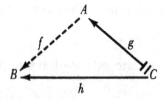

C into B. We are to construct an extension f that maps A into B. The idea of the proof is to extend step by step, transfinitely; a typical step is of the kind described in the extension lemma. An efficient way to carry out the transfinite process is, as usual, by Zorn's lemma. In view of Zorn's lemma we may and do assume that h is a maximal homomorphism from C to B, that is, that it admits of no extension to a subalgebra of A that is strictly larger than C.

If $C \neq A$, then A has an element, r say, that is not in C. We derive a contradiction by extending h to the algebra generated by C and r; by an obvious change of notation we may and do assume that that generated algebra is A itself. Let p_* be the supremum in B of all the elements of the form $h(s)$, where $s \in C$ and $s \leq r$; similarly, let p^* be the infimum in B of all the elements of the form $h(t)$, where $t \in C$ and $r \leq t$. (Here is where the completeness of B is used.) If $s \leq r \leq t$, then $h(s) \leq h(t)$; this implies that $p_* \leq p^*$, and hence that there exists an element p in B such that $p_* \leq p \leq p^*$. The expected contradiction is now a direct consequence of the extension lemma.

COROLLARY. *A Boolean space is injective if and only if it is complete.*

From the characterization of injectivity now at hand we can deduce for injective algebras the results obtained in the preceding section for projective algebras.

COROLLARY 1. *A Boolean algebra is injective if and only if it is a retract of a complete algebra.*

COROLLARY 2. *A Boolean algebra is injective if and only if it is an absolute subretract.*

COROLLARY 3. *A Boolean space is projective if and only if it is a retract of a complete space.*

COROLLARY 4. *A Boolean space is projective if and only if it is an absolute quotient retract.*

EPILOGUE

There is much more to Boolean algebras than is covered in this volume. The reader who wants to learn more should consult Sikorski's scholarly book (Berlin, 1960) and its excellent bibliography.

Index

A CATALOG OF SELECTED
DOVER BOOKS
IN SCIENCE AND MATHEMATICS

Mathematics-Bestsellers

HANDBOOK OF MATHEMATICAL FUNCTIONS: with Formulas, Graphs, and Mathematical Tables, Edited by Milton Abramowitz and Irene A. Stegun. A classic resource for working with special functions, standard trig, and exponential logarithmic definitions and extensions, it features 29 sets of tables, some to as high as 20 places. 1046pp. 8 x 10 1/2. 0-486-61272-4

ABSTRACT AND CONCRETE CATEGORIES: The Joy of Cats, Jiri Adamek, Horst Herrlich, and George E. Strecker. This up-to-date introductory treatment employs category theory to explore the theory of structures. Its unique approach stresses concrete categories and presents a systematic view of factorization structures. Numerous examples. 1990 edition, updated 2004. 528pp. 6 1/8 x 9 1/4. 0-486-46934-4

MATHEMATICS: Its Content, Methods and Meaning, A. D. Aleksandrov, A. N. Kolmogorov, and M. A. Lavrent'ev. Major survey offers comprehensive, coherent discussions of analytic geometry, algebra, differential equations, calculus of variations, functions of a complex variable, prime numbers, linear and non-Euclidean geometry, topology, functional analysis, more. 1963 edition. 1120pp. 5 3/8 x 8 1/2. 0-486-40916-3

INTRODUCTION TO VECTORS AND TENSORS: Second Edition--Two Volumes Bound as One, Ray M. Bowen and C.-C. Wang. Convenient single-volume compilation of two texts offers both introduction and in-depth survey. Geared toward engineering and science students rather than mathematicians, it focuses on physics and engineering applications. 1976 edition. 560pp. 6 1/2 x 9 1/4. 0-486-46914-X

AN INTRODUCTION TO ORTHOGONAL POLYNOMIALS, Theodore S. Chihara. Concise introduction covers general elementary theory, including the representation theorem and distribution functions, continued fractions and chain sequences, the recurrence formula, special functions, and some specific systems. 1978 edition. 272pp. 5 3/8 x 8 1/2. 0-486-47929-3

ADVANCED MATHEMATICS FOR ENGINEERS AND SCIENTISTS, Paul DuChateau. This primary text and supplemental reference focuses on linear algebra, calculus, and ordinary differential equations. Additional topics include partial differential equations and approximation methods. Includes solved problems. 1992 edition. 400pp. 7 1/2 x 9 1/4. 0-486-47930-7

PARTIAL DIFFERENTIAL EQUATIONS FOR SCIENTISTS AND ENGINEERS, Stanley J. Farlow. Practical text shows how to formulate and solve partial differential equations. Coverage of diffusion-type problems, hyperbolic-type problems, elliptic-type problems, numerical and approximate methods. Solution guide available upon request. 1982 edition. 414pp. 6 1/8 x 9 1/4. 0-486-67620-X

VARIATIONAL PRINCIPLES AND FREE-BOUNDARY PROBLEMS, Avner Friedman. Advanced graduate-level text examines variational methods in partial differential equations and illustrates their applications to free-boundary problems. Features detailed statements of standard theory of elliptic and parabolic operators. 1982 edition. 720pp. 6 1/8 x 9 1/4. 0-486-47853-X

LINEAR ANALYSIS AND REPRESENTATION THEORY, Steven A. Gaal. Unified treatment covers topics from the theory of operators and operator algebras on Hilbert spaces; integration and representation theory for topological groups; and the theory of Lie algebras, Lie groups, and transform groups. 1973 edition. 704pp. 6 1/8 x 9 1/4. 0-486-47851-3

Browse over 9,000 books at www.doverpublications.com

CATALOG OF DOVER BOOKS

A SURVEY OF INDUSTRIAL MATHEMATICS, Charles R. MacCluer. Students learn how to solve problems they'll encounter in their professional lives with this concise single-volume treatment. It employs MATLAB and other strategies to explore typical industrial problems. 2000 edition. 384pp. 5 3/8 x 8 1/2. 0-486-47702-9

NUMBER SYSTEMS AND THE FOUNDATIONS OF ANALYSIS, Elliott Mendelson. Geared toward undergraduate and beginning graduate students, this study explores natural numbers, integers, rational numbers, real numbers, and complex numbers. Numerous exercises and appendixes supplement the text. 1973 edition. 368pp. 5 3/8 x 8 1/2. 0-486-45792-3

A FIRST LOOK AT NUMERICAL FUNCTIONAL ANALYSIS, W. W. Sawyer. Text by renowned educator shows how problems in numerical analysis lead to concepts of functional analysis. Topics include Banach and Hilbert spaces, contraction mappings, convergence, differentiation and integration, and Euclidean space. 1978 edition. 208pp. 5 3/8 x 8 1/2. 0-486-47882-3

FRACTALS, CHAOS, POWER LAWS: Minutes from an Infinite Paradise, Manfred Schroeder. A fascinating exploration of the connections between chaos theory, physics, biology, and mathematics, this book abounds in award-winning computer graphics, optical illusions, and games that clarify memorable insights into self-similarity. 1992 edition. 448pp. 6 1/8 x 9 1/4. 0-486-47204-3

SET THEORY AND THE CONTINUUM PROBLEM, Raymond M. Smullyan and Melvin Fitting. A lucid, elegant, and complete survey of set theory, this three-part treatment explores axiomatic set theory, the consistency of the continuum hypothesis, and forcing and independence results. 1996 edition. 336pp. 6 x 9. 0-486-47484-4

DYNAMICAL SYSTEMS, Shlomo Sternberg. A pioneer in the field of dynamical systems discusses one-dimensional dynamics, differential equations, random walks, iterated function systems, symbolic dynamics, and Markov chains. Supplementary materials include PowerPoint slides and MATLAB exercises. 2010 edition. 272pp. 6 1/8 x 9 1/4. 0-486-47705-3

ORDINARY DIFFERENTIAL EQUATIONS, Morris Tenenbaum and Harry Pollard. Skillfully organized introductory text examines origin of differential equations, then defines basic terms and outlines general solution of a differential equation. Explores integrating factors; dilution and accretion problems; Laplace Transforms; Newton's Interpolation Formulas, more. 818pp. 5 3/8 x 8 1/2. 0-486-64940-7

MATROID THEORY, D. J. A. Welsh. Text by a noted expert describes standard examples and investigation results, using elementary proofs to develop basic matroid properties before advancing to a more sophisticated treatment. Includes numerous exercises. 1976 edition. 448pp. 5 3/8 x 8 1/2. 0-486-47439-9

THE CONCEPT OF A RIEMANN SURFACE, Hermann Weyl. This classic on the general history of functions combines function theory and geometry, forming the basis of the modern approach to analysis, geometry, and topology. 1955 edition. 208pp. 5 3/8 x 8 1/2. 0-486-47004-0

THE LAPLACE TRANSFORM, David Vernon Widder. This volume focuses on the Laplace and Stieltjes transforms, offering a highly theoretical treatment. Topics include fundamental formulas, the moment problem, monotonic functions, and Tauberian theorems. 1941 edition. 416pp. 5 3/8 x 8 1/2. 0-486-47755-X

Browse over 9,000 books at www.doverpublications.com

Mathematics–Logic and Problem Solving

PERPLEXING PUZZLES AND TANTALIZING TEASERS, Martin Gardner. Ninety-three riddles, mazes, illusions, tricky questions, word and picture puzzles, and other challenges offer hours of entertainment for youngsters. Filled with rib-tickling drawings. Solutions. 224pp. 5 3/8 x 8 1/2. 0-486-25637-5

MY BEST MATHEMATICAL AND LOGIC PUZZLES, Martin Gardner. The noted expert selects 70 of his favorite "short" puzzles. Includes The Returning Explorer, The Mutilated Chessboard, Scrambled Box Tops, and dozens more. Complete solutions included. 96pp. 5 3/8 x 8 1/2. 0-486-28152-3

THE LADY OR THE TIGER?: and Other Logic Puzzles, Raymond M. Smullyan. Created by a renowned puzzle master, these whimsically themed challenges involve paradoxes about probability, time, and change; metapuzzles; and self-referentiality. Nineteen chapters advance in difficulty from relatively simple to highly complex. 1982 edition. 240pp. 5 3/8 x 8 1/2. 0-486-47027-X

SATAN, CANTOR AND INFINITY: Mind-Boggling Puzzles, Raymond M. Smullyan. A renowned mathematician tells stories of knights and knaves in an entertaining look at the logical precepts behind infinity, probability, time, and change. Requires a strong background in mathematics. Complete solutions. 288pp. 5 3/8 x 8 1/2.

0-486-47036-9

THE RED BOOK OF MATHEMATICAL PROBLEMS, Kenneth S. Williams and Kenneth Hardy. Handy compilation of 100 practice problems, hints and solutions indispensable for students preparing for the William Lowell Putnam and other mathematical competitions. Preface to the First Edition. Sources. 1988 edition. 192pp. 5 3/8 x 8 1/2. 0-486-69415-1

KING ARTHUR IN SEARCH OF HIS DOG AND OTHER CURIOUS PUZZLES, Raymond M. Smullyan. This fanciful, original collection for readers of all ages features arithmetic puzzles, logic problems related to crime detection, and logic and arithmetic puzzles involving King Arthur and his Dogs of the Round Table. 160pp. 5 3/8 x 8 1/2.

0-486-47435-6

UNDECIDABLE THEORIES: Studies in Logic and the Foundation of Mathematics, Alfred Tarski in collaboration with Andrzej Mostowski and Raphael M. Robinson. This well-known book by the famed logician consists of three treatises: "A General Method in Proofs of Undecidability," "Undecidability and Essential Undecidability in Mathematics," and "Undecidability of the Elementary Theory of Groups." 1953 edition. 112pp. 5 3/8 x 8 1/2. 0-486-47703-7

LOGIC FOR MATHEMATICIANS, J. Barkley Rosser. Examination of essential topics and theorems assumes no background in logic. "Undoubtedly a major addition to the literature of mathematical logic." – *Bulletin of the American Mathematical Society*. 1978 edition. 592pp. 6 1/8 x 9 1/4. 0-486-46898-4

INTRODUCTION TO PROOF IN ABSTRACT MATHEMATICS, Andrew Wohlgemuth. This undergraduate text teaches students what constitutes an acceptable proof, and it develops their ability to do proofs of routine problems as well as those requiring creative insights. 1990 edition. 384pp. 6 1/2 x 9 1/4. 0-486-47854-8

FIRST COURSE IN MATHEMATICAL LOGIC, Patrick Suppes and Shirley Hill. Rigorous introduction is simple enough in presentation and context for wide range of students. Symbolizing sentences; logical inference; truth and validity; truth tables; terms, predicates, universal quantifiers; universal specification and laws of identity; more. 288pp. 5 3/8 x 8 1/2. 0-486-42259-3

Mathematics–Algebra and Calculus

VECTOR CALCULUS, Peter Baxandall and Hans Liebeck. This introductory text offers a rigorous, comprehensive treatment. Classical theorems of vector calculus are amply illustrated with figures, worked examples, physical applications, and exercises with hints and answers. 1986 edition. 560pp. 5 3/8 x 8 1/2.   0-486-46620-5

ADVANCED CALCULUS: An Introduction to Classical Analysis, Louis Brand. A course in analysis that focuses on the functions of a real variable, this text introduces the basic concepts in their simplest setting and illustrates its teachings with numerous examples, theorems, and proofs. 1955 edition. 592pp. 5 3/8 x 8 1/2.   0-486-44548-8

ADVANCED CALCULUS, Avner Friedman. Intended for students who have already completed a one-year course in elementary calculus, this two-part treatment advances from functions of one variable to those of several variables. Solutions. 1971 edition. 432pp. 5 3/8 x 8 1/2.   0-486-45795-8

METHODS OF MATHEMATICS APPLIED TO CALCULUS, PROBABILITY, AND STATISTICS, Richard W. Hamming. This 4-part treatment begins with algebra and analytic geometry and proceeds to an exploration of the calculus of algebraic functions and transcendental functions and applications. 1985 edition. Includes 310 figures and 18 tables. 880pp. 6 1/2 x 9 1/4.   0-486-43945-3

BASIC ALGEBRA I: Second Edition, Nathan Jacobson. A classic text and standard reference for a generation, this volume covers all undergraduate algebra topics, including groups, rings, modules, Galois theory, polynomials, linear algebra, and associative algebra. 1985 edition. 528pp. 6 1/8 x 9 1/4.   0-486-47189-6

BASIC ALGEBRA II: Second Edition, Nathan Jacobson. This classic text and standard reference comprises all subjects of a first-year graduate-level course, including in-depth coverage of groups and polynomials and extensive use of categories and functors. 1989 edition. 704pp. 6 1/8 x 9 1/4.   0-486-47187-X

CALCULUS: An Intuitive and Physical Approach (Second Edition), Morris Kline. Application-oriented introduction relates the subject as closely as possible to science with explorations of the derivative; differentiation and integration of the powers of x; theorems on differentiation, antidifferentiation; the chain rule; trigonometric functions; more. Examples. 1967 edition. 960pp. 6 1/2 x 9 1/4.   0-486-40453-6

ABSTRACT ALGEBRA AND SOLUTION BY RADICALS, John E. Maxfield and Margaret W. Maxfield. Accessible advanced undergraduate-level text starts with groups, rings, fields, and polynomials and advances to Galois theory, radicals and roots of unity, and solution by radicals. Numerous examples, illustrations, exercises, appendixes. 1971 edition. 224pp. 6 1/8 x 9 1/4.   0-486-47723-1

AN INTRODUCTION TO THE THEORY OF LINEAR SPACES, Georgi E. Shilov. Translated by Richard A. Silverman. Introductory treatment offers a clear exposition of algebra, geometry, and analysis as parts of an integrated whole rather than separate subjects. Numerous examples illustrate many different fields, and problems include hints or answers. 1961 edition. 320pp. 5 3/8 x 8 1/2.   0-486-63070-6

LINEAR ALGEBRA, Georgi E. Shilov. Covers determinants, linear spaces, systems of linear equations, linear functions of a vector argument, coordinate transformations, the canonical form of the matrix of a linear operator, bilinear and quadratic forms, and more. 387pp. 5 3/8 x 8 1/2.   0-486-63518-X

Mathematics–Probability and Statistics

BASIC PROBABILITY THEORY, Robert B. Ash. This text emphasizes the probabilistic way of thinking, rather than measure-theoretic concepts. Geared toward advanced undergraduates and graduate students, it features solutions to some of the problems. 1970 edition. 352pp. 5 3/8 x 8 1/2. 0-486-46628-0

PRINCIPLES OF STATISTICS, M. G. Bulmer. Concise description of classical statistics, from basic dice probabilities to modern regression analysis. Equal stress on theory and applications. Moderate difficulty; only basic calculus required. Includes problems with answers. 252pp. 5 5/8 x 8 1/4. 0-486-63760-3

OUTLINE OF BASIC STATISTICS: Dictionary and Formulas, John E. Freund and Frank J. Williams. Handy guide includes a 70-page outline of essential statistical formulas covering grouped and ungrouped data, finite populations, probability, and more, plus over 1,000 clear, concise definitions of statistical terms. 1966 edition. 208pp. 5 3/8 x 8 1/2. 0-486-47769-X

GOOD THINKING: The Foundations of Probability and Its Applications, Irving J. Good. This in-depth treatment of probability theory by a famous British statistician explores Keynesian principles and surveys such topics as Bayesian rationality, corroboration, hypothesis testing, and mathematical tools for induction and simplicity. 1983 edition. 352pp. 5 3/8 x 8 1/2. 0-486-47438-0

INTRODUCTION TO PROBABILITY THEORY WITH CONTEMPORARY APPLICATIONS, Lester L. Helms. Extensive discussions and clear examples, written in plain language, expose students to the rules and methods of probability. Exercises foster problem-solving skills, and all problems feature step-by-step solutions. 1997 edition. 368pp. 6 1/2 x 9 1/4. 0-486-47418-6

CHANCE, LUCK, AND STATISTICS, Horace C. Levinson. In simple, non-technical language, this volume explores the fundamentals governing chance and applies them to sports, government, and business. "Clear and lively ... remarkably accurate." – *Scientific Monthly*. 384pp. 5 3/8 x 8 1/2. 0-486-41997-5

FIFTY CHALLENGING PROBLEMS IN PROBABILITY WITH SOLUTIONS, Frederick Mosteller. Remarkable puzzlers, graded in difficulty, illustrate elementary and advanced aspects of probability. These problems were selected for originality, general interest, or because they demonstrate valuable techniques. Also includes detailed solutions. 88pp. 5 3/8 x 8 1/2. 0-486-65355-2

EXPERIMENTAL STATISTICS, Mary Gibbons Natrella. A handbook for those seeking engineering information and quantitative data for designing, developing, constructing, and testing equipment. Covers the planning of experiments, the analyzing of extreme-value data; and more. 1966 edition. Index. Includes 52 figures and 76 tables. 560pp. 8 3/8 x 11. 0-486-43937-2

STOCHASTIC MODELING: Analysis and Simulation, Barry L. Nelson. Coherent introduction to techniques also offers a guide to the mathematical, numerical, and simulation tools of systems analysis. Includes formulation of models, analysis, and interpretation of results. 1995 edition. 336pp. 6 1/8 x 9 1/4. 0-486-47770-3

INTRODUCTION TO BIOSTATISTICS: Second Edition, Robert R. Sokal and F. James Rohlf. Suitable for undergraduates with a minimal background in mathematics, this introduction ranges from descriptive statistics to fundamental distributions and the testing of hypotheses. Includes numerous worked-out problems and examples. 1987 edition. 384pp. 6 1/8 x 9 1/4. 0-486-46961-1